M000266609

The Greatest Story
NEVER Told

The Assured Triumph of Human Inevitability and Superiority

By

Elvis Newman

Know the real cause of your Post Traumatic Slave Syndrome
Deny Ignorance, Propaganda and Invisible Prison Cell
Deny deception of Corpenican Proportions

Strategic Book Publishing and Rights Co.

Copyright © 2014 Elvis Newman. All rights reserved.

No part of this book may be reproduced or transmitted in any form or by any means, graphic, electronic, or mechanical, including photocopying, recording, taping, or by any information storage retrieval system, without the permission, in writing, of the publisher. For more information, send a letter to our Houston, TX address, Attention Subsidiary Rights Department, or email: support@sbpra.net.

Strategic Book Publishing and Rights Co.
12620 FM 1960, Suite A4-507
Houston, TX 77065
www.sbpra.com

For information about special discounts for bulk purchases, please contact Strategic Book Publishing and Rights Co. Special Sales, at bookorder@sbpra.net.

ISBN: 978-1-62857-124-0

It is clear that the Sumerian clay tablets study will continue to interest areas of science, history, archaeology, cosmology, etc. due to its continued "effect" of encompassing studies from unrelated fields. What might be the beauty of Sitchin's theory is that it gives people, scientists, and scholars the confidence to create new theories and discover new findings, even though the final and conclusive proof may still be unknown or unaccepted by the populace or scientific mainstream.

"Once you eliminate the impossible, whatever remains, no matter how improbable, must be the truth."

-Arthur Conan Doyle, Sr. quotes

Zecharia Sitchin was an Azerbaijani born American author of books proposing an explanation for human origins involving ancient astronauts.

Born: July 11, 1920
Died: October 9, 2010, New York

Contents

Introduction

It is my sincere hope that everyone who reads this work will be inspired to question things and to search out these and other new truths and discoveries for themselves.

We should all be engaged in the most important intellectual dialog, exchange, and sharing of the twenty-first century.

I do not ask or expect anyone to blindly believe what is written within the pages of this book without investigating all the evidence for themselves. In the quest for truth, all our shutters will be opened onto a brave new world, as if seeing with new eyes life and society in all its splendor and glory with an ever greater sensitivity and realization.

Remember, with Truth, sometimes it is stranger than fiction, sometimes it is vastly outnumbered, sometimes it is out there somewhere, and sometimes it lies in our inner awakening.

I hope the investigations, interchanges, and interactions that surround the greatest discovery and controversy that pertain to the findings in the Sumerian clay tablets will eventually unite us in a beautiful harmony called love. In our quest for the ultimate Truth, we shall progress and usher in the Golden Age.

I thank you in advance for taking time to read my book and using it as a thought-provoking work.

Part 1

Lost Book of Enki

Part 1 is a rewritten account of The Lost Book of Enki by Zecharia Sitchin

Some four hundred and forty-five thousand years ago, humanoid aliens called Anunnakis came to Earth to mine gold to heal their atmosphere and save their planet, Nibiru. To this end, they built a full-fledged Mission Earth—with a mission control center, a spaceport, mining operations, and even a way station on Mars. King Anu and the Council on Nibiru were desperate for more miners and decreed their eminent geneticist, Enki, to employ Anunnakis' advance genetic engineering to fashion a race of primitive workers. Through a series of blunders, the perfect slave race, intelligent and subservient prototype, Homo sapiens, was created.

The Homo sapiens multiplied in great numbers. Enhanced hybrids were begotten through gradual intermarrying of the descendants of Anunnakis and humans. The deluge in Noah's times that catastrophically swept over the Earth wiped out most of the settlements. The Anunnakis posed themselves as gods, granting mankind civilization, social hierarchy, domesticated animals, and plants. They also taught mankind astronomy, mathematics, science, laws, architecture, metallurgy, agriculture etc, etc . . . Then about four thousand years ago, internal rivalries among the Anunnaki prompted different factions in the Anunnakis to nuke each other. Most of them returned to Nibiru after the event.

The planet Nibiru is a big planet, sultry and vivacious. It looks like a massive brown dwarf star, which glows in the infrared band, and would have appeared to be on fire to ancient Earth witnesses. Incapacitated, Earth scientists have concluded that there could be no life on Nibiru. But there is life on Nibiru—life that should resemble ours, but doesn't. At a time of the darkest moment in their history, the inhabitants of Nibiru, the Anunnaki, led desperate lives, suffering from great and terrible need. They began to search the universe in an effort to gratify that need. They sought a planet on which life is healthy, vibrant, and strong. They needed finest gold particles broken down and manufactured with nanotechnology to heal their atmosphere. They needed miners to do their work, to labor and slave for them, to manufacture their golden dreams. The Anunnaki needed slaves, and they had chosen the planet on which their slaves would be created after carrying out their sophisticated and high-tech genetic experiment and foolproof feasibility study. Part 1 is the story of their presence on Earth around B.C. 450,100.

Chapter One

The Debate for the Anunnaki and Ape Hybrids

B.C. 450,100.

In the Court of King Anu, Planet Nibiru, the important issue of primitive workers on Colony Earth was raised. All the elders, wise men, savants, gifted, prodigies, clairvoyants, famous, scientists, leaders, commanders, explorers, entrepreneurs, and people's representatives were gathered in the assembly for a long and bitter related discussion on morality, labor laws, enslavement mandates, interplanetary journey stipulations, and Planet Nibiru's atmospheric conditions.

The central issue was: Should primitive workers be genetically engineered and created to solve the astronauts' labor dispute and possible mutiny?

Major proponents were led by Enki, second commander of Terran Operations.

"Make Homo erectus more intelligent by changing them into Homo sapiens. Our commands he will then understand. Our tools he will then handle. He shall perform the toil in the excavations. We will have no more labor disputes coming from our astronauts working as gold miners."

"It is not a new creature created by us. The plan is to give this ape creature that we found on Earth more ability, make them more to our likeness. Only a tiny segment of our DNA is all that is needed."

Opponents were led by Enlil, supreme commander of Terran Operations.

"Technological improvements in production methods, toolmaking, and machinery would make a better choice. Anunnaki should not play God by creating slave beings!"

"Creation is the power held by the Father of All Beginnings, not Anunnaki. We are sent to Earth to obtain gold to repair Nibiru's dwindling atmosphere, not to replace the Father of All Beginnings."

There was no verdict after many days of intense debate and deliberation. On the seventh day, the Supreme Council of King Anu reconvened for the seventh time.

"Can we get the finest gold dusts in the galaxy in any other way? Our Planet Nibiru is dying. The survival of the glorious Anunnaki civilization hangs in the balance. We must get gold to repair Nibiru's broken ozone. We must get the finest gold dusts in the galaxy at all costs."

"Let us forget about interplanetary travel rules and save Nibiru. If there is no Planet Nibiru, there will be no Anunnaki civilization. If there is no Anunnaki civilization, there is no need for interplanetary travel rules."

"Let us save Nibiru by the creation of slave workers through high-tech genetic engineering!"

Chapter Two

The Story of the Garden of Eden,
Prehistoric Iraq

B.C. 450,100. Enki, second commander of Terran Operations, was rocketed to Earth with fifty medics, six hundred miners, three hundred astronauts, and a hundred administrators.

So the Anunnaki came to the Earth around four hundred and fifty thousand years ago to mine gold in what is now Africa. The main mining center was in today's Zimbabwe, an area the Sumerians called AB.ZU (deep deposit). The gold mined by the Anunnaki was shipped back to their home planet from bases in the Middle East.

At first, the gold mining was done by the intelligent and physically robust Anunnaki astronauts and miners. These men were of the respectable elite back home. However, they were made to toil and work as miners on Earth in order to save Nibiru. Eventually, there were riots and rebellions, and the Anunnaki royal leadership decided to create a new slave race to do the work.

Chapter Three

Terran Wildlife Natural Reserve and Research Center

Enki and his son Ningizidda led a team of biologists to set up a lab for Earth animal studies. Special interest was given especially to Homo erectus, a species Enki identified as due to evolve in a few million years into Homo sapiens, the species like the Nibirans.

The apes kept at the reserve were studied carefully by the Anunnaki. Found in the African forests, the apes walked erect on two legs. Their forelegs they used as arms. They lived among the animals of the steppe. They ate plants with their mouths. They drank water from lakes and ditches. It was a thrill to see them alive! They were kept in strong cages. At the sight of visitors, they jumped up with fists beating on the cage bars, grunting and snorting and speaking no words. Just like the Anunnaki, they had males and females!

The study of Homo erectus revealed that just like the chimpanzees, gorillas, and orangutans, they had twenty-four pairs of chromosomes. Ningishzidda, president of the Biogenetics Council, explained to Ninmah, the chief medical officer, how the second and third chromosomes of Homo erectus could be fused together to give rise to the intelligent new ape. This fusion could only be accomplished in the Terran genetic splicing laboratory through manipulation of the egg. No disease or natural genetic condition had ever caused a species to alter its chromosomal structure!

(*All members of Hominidae except humans have twenty-four pairs of chromosomes. Humans have only twenty-three pairs of chromosomes. Human chromosome two is widely accepted to be a result of an end-to-end fusion of two ancestral chromosomes. Wikipedia.)

To the astonished Ninmah, Ningishzidda continued, "There are things that we can do now that surpass your wildest dreams. We can jump the gun on the new ape's progress, and make it into Homo sapiens through infusion of some Anunnaki gene segments!"

"Oh, Homo erectus! They are too odd, too monstrous, and too barbaric!"

"That is why you are needed for their perfection into Homo sapiens!"

Enki, second commander of Terran Operations, said to Ninmah and Ningishzidda, "Gather your teams and start working!"

Footnote

In Inside the Human Genome: A Case for Non-Intelligent Design by John C. Avise, the author argues that from the perspectives of biochemistry and molecular genetics there are numerous flaws that exist in the biological world. There is overwhelming scientific evidence for genomic imperfection, imperfections that extend deep down to the level of our genes, and evidence also exists that there are many gross deficiencies in human DNA, ranging from mutational defects to built-in design faults.

The oversimplified and often touted claim that the DNA of chimpanzees (Pan troglodytes) and humans (Homo sapiens) are about 98 percent similar, a claim made by Oxford professor Richard Dawkins and other evolution proponents, can be refuted by genomic scientists. Dr. Jeffrey Tomkins explored how chimps became our closest ancestor and the DNA based early evidence that started this misconception in his book, *More Than a Monkey.*

Chapter Four

A.C.T.G Organization / Adenine, Cytosine, Thymine and Guanine

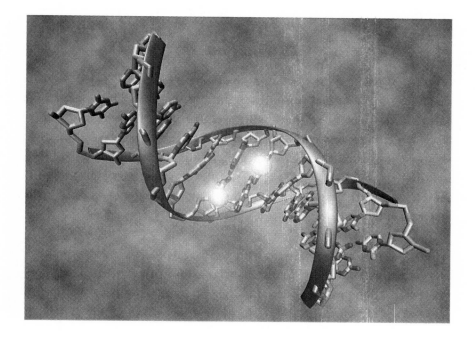

Anunnaki Center for Terran Genetics
Garden of Eden, Prehistoric Iraq

Anunnaki Human Hybrid Research Commission

We have the strongest hybrids management systems in the solar system.
Thanks to our xenobiologists' hard work, we now have made the most

recent accomplishment in hybrid gene and DNA formation. Human genetics is an important milestone in our efforts to secure profitable and healthy primitive workers. To make sure that our research and science remain topnotch, new legislation and King Anu's appropriations for science and management are critical to maintain this momentum.

Because what we have done so far is historic, we shall commemorate our achievements by encoding the intelligent ape's genetic information as a sequence of nucleotides: adenine, thymine, cytosine, and guanine, with the acronym for our organization A.C.T.G.

Footnote

A group of researchers working at the Human Genome Project indicate that they made an astonishing scientific discovery: They believe the so-called 97 percent of noncoding sequences in human DNA is no less than genetic code of extraterrestrial life forms, says Prof. Sam Chang, the group leader. The overwhelming majority of human DNA is "off-world" in origin. The apparent "extraterrestrial junk genes" merely "enjoy the ride" with hardworking active genes, passed from generation to generation.

After comprehensive analysis with the assistance of other scientists, computer programmers, mathematicians, and other learned scholars, Professor Chang had wondered if the apparently "junk human DNA" was created by some kind of "extraterrestrial programmer." The alien chunks within human DNA, Professor Chang further observes, "have its own veins, arteries, and its own immune system that vigorously resists all our anticancer drugs."

Professor Chang further stipulates, "Our hypothesis is that a higher extraterrestrial life-form was engaged in creating new life and planting it on various planets. Earth is just one of them. Perhaps, after programming, our creators grow us the same way we grow bacteria in

Petri dishes. We can't know their motives—whether it was a scientific experiment, or a way of preparing new planets for colonization, or is it long time ongoing business of seedling life in the universe."

Professor Chang and his research colleagues show that apparent "extraterrestrial programming" gaps in DNA sequencing, precipitated by a hypothesized rush to create human life on Earth, presented humankind with illogical growth of mass cells we know as cancer. Very likely in an apparent rush, the "extraterrestrial programmers" may have cut down drastically on big code and delivered basic program intended for Earth.

Professor Chang further indicates, "What we see in our DNA is a program consisting of two versions—a big code and basic code." Mr. Chang then affirms that the "first fact is the complete 'program' was positively not written on Earth; that is now a verified fact. The second fact is that genes by themselves are not enough to explain evolution; there must be something more in 'the game.'"

"Sooner or later," Professor Chang says, "we have to come to grips with the unbelievable notion that every life on Earth carries genetic code for his extraterrestrial cousin and that evolution is not what we think it is." (John Stokes. *Scientists Find Extraterrestrial Genes In Human DNA: Do civilizations of advanced human beings exist scattered in the Galaxy?* 1-11-7. http://rense.com/general74/d3af.htm)

Chapter Five

Natural Cycle in
Vitro Fertilization (IVF)

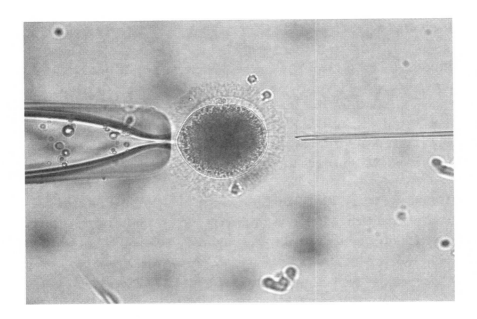

Series and series of genetic experiments were carried out fervently. And their results and products were carefully documented.

Admixture of essences was prepared with mitochondria extracts, Anunnaki spliced genes segments, and ova of female Homo erectus impregnated with Anunnaki sperm.

Fertilized ova were inserted back to the womb of the ape Homo erectus.

A birth was expected. No birth was forthcoming.

In desperation, Ninmah made a dissection and pulled out the baby with tongs. It was a living being.

Ningishzidda cried out, "We are successful!"

Ninmah was not filled with joy.

The newborn was shaggy with hair all over. His foreparts resembled the Homo erectus. His hind parts resembled the Anunnaki.

The newborn was breastfed by the mother Homo erectus. The newborn was growing fast. A day in the Anunnaki planet is about a month in Africa.

The child grew tall, not looking like the Anunnaki.

His hands could not hold tools.

He could only make grunting sounds instead of speech.

Intravaginal culture (IVC)

Once again, the experiment was done after changing the admixture. Nimah carefully considered and examined the Anunnaki essences. One bit she took from one. One bit she took from another.

In the Nibiru made test tube, the ova of an Earth female was inseminated. The tube was then hermetically closed and was placed in the maternal vagina and held by a diaphragm for incubation for forty-four to fifty hours.

After this time, the content of the tube was examined, and embryos were transferred to the uterus.

There was conception. There was birth giving. This one looked more like the Anunnaki. The baby was breastfed by his ape mother and nurtured.

He looked appealing, hands fit to hold a tool. However, he was lacking in senses. He could not hear and had bad eyesight.

The persistent and patient Ninmah repeated the experiments with varying compositions of the admixtures thousands of times.

Chapter Six

Nimah's Progress Entry

"Hello there, I am Nimah, the chief medical officer for the Terran Genome Project, and I am here to present to you the creatures that will change the world as we know it—the humans, our slave race.

"This creature can manipulate tools and create different things. For example, give it an object and ask it to turn it into something. It will comply. Most importantly, if you ask it, it will work as a miner in our excavations. This creature also holds many wonderful surprises as promised. My team and I will keep trying and trying until the perfect creature as envisioned by our leaders is created."

Hundreds of babies were created each year and their conditions compared. The dead babies were then dissected and their organs and parts comparatively analyzed.

The best case of each year was selected for report. Even then, the best cases were often fraught with defects.

Enki, Ninmah, and Ningishzidda mapped chromosomes, genes, and genomes. They fertilized ova in test-tube flasks with sperm soaked in Nibiran blood serum and mineral nutrients. They experimented with cloning, cell fusion, and recombinant technology—cutting DNA strands with enzymes and targeted viruses. They absorbed sperm in genetic material to be used for fertilization. They spliced DNA patches of other species to create, at first, hybrids, which were unable to reproduce. Then Ningishzidda isolated the XX and XY chromosomes, which allowed the creation of fertile Nibiran/erectus mine slaves [ZS, 1990, Genesis: 158 - 182, 202].

Progress entry 30091 B.C. 450,000: Using ultrasound fetal scanning, we found this new baby to have paralyzed feet.

Progress entry 48762 B.C. 449,999: The new being had semen dripping.

Progress entry 58973 B.C. 449,998: This new one had progeria, which caused trembling hands.

Progress entry 67554 B.C. 449,997: Amniocentesis indicated this new one had a malfunctioning liver.

Progress entry 76545 B.C. 449,996: Upon birth, we found this new one had hands too short to reach the mouth.

Progress entry 87656 B.C. 449,995: Through fetal electrocardiogram (ECG), we could tell this new one had alveoli atrophy in the lungs, and he could not breathe enough air.

Enki was greatly disappointed.

"We just cannot create a primitive worker!"

Nimah said to Ningishzidda and Enki, "By trial and error, I am realizing what works. We must persist."

Chapter Seven

Ningishzidda's Junk DNA and Abnormalities

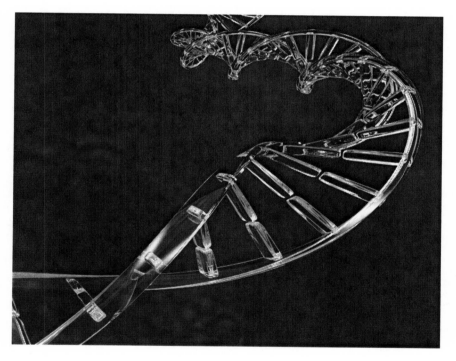

Another try. Another deformed creature was made.

It was very late, and everyone had left the laboratory except Ningishzidda. In daytime, the street was dusty from excavations in the mines, but at night, the dust and dew settled. Now at night, it was quiet, and he could think alone. While he was brilliant in his field, the enigma of life was to him still the greatest mystery. Only the moon kept him company.

Another try. Another deformed creature was made.

Several weeks later, Ningishzidda again stayed back after everyone had left the laboratory.

"Junk DNA sequence. Maybe the junk DNA sequence is the problem." He kept trying and trying, until his eyes could not tell all the junk DNA materials from the junk garbage. And he fell asleep at the table.

Another try. Another deformed creature was made.

Like a trial, you don't have to be there to know the verdict. Ningishzidda knew all the time what it would be. Still, Ningishzidda tried again.

Another try. Another deformed creature was made.

Footnote

Four thousand defects in the human genome:

Four thousand defects in the human genome are carried in everyone's DNA, more than any other species, and they unnaturally survived in the human gene pool instead of bred out. Over four thousand human diseases are caused by single gene defects. Single gene disorders can be passed on to subsequent generations in several ways. Source: http://en.wikipedia.org/wiki/Genetic_disorder

Humans have only twenty-three pairs of chromosomes

All members of Hominidae except humans have twenty-four pairs of chromosomes. Humans have only twenty-three pairs of chromosomes.

Human chromosome two is widely accepted to be a result of an end-to-end fusion of two ancestral chromosomes. The evidence for this includes:

1. The correspondence of chromosome two to two ape chromosomes. The closest human relative, the chimpanzee, has nearly identical DNA sequences to human chromosome two, but they are found in two separate chromosomes. The same is true of the more distant gorilla and orangutan.

2. The presence of a vestigial centromere. Normally, a chromosome has just one centromere, but in chromosome two there are remnants of a second centromere.

3. The presence of vestigial telomeres. These are normally found only at the ends of a chromosome, but in chromosome two, there are additional telomere sequences in the middle.

   Chromosome two presents very strong evidence in favor of the common descent of humans and other apes. According to researcher J. W. IJdo, "We conclude that the locus cloned in cosmids c8.1 and c29B is the relic of an ancient telomere-telomere fusion and marks the point at which two ancestral ape chromosomes fused to give rise to human chromosome 2."

Source: http://en.wikipedia.org/wiki/Chromosome_2_(human)

1. Robertsonian Translocations "People with Robertsonian translocations have only forty-five chromosomes in each of their cells, yet all essential genetic material is present, and they appear normal. Their children, however, may either be normal and carry the fusion chromosome (depending which chromosome is represented in the gamete), or they may inherit a missing or extra long arm of an acrocentric chromosome." Kenneth F. Trofatter, Jr., MD, PhD.

2. The common chimp (Pan troglodytes) and human Y chromosomes

The common chimp (Pan troglodytes) and human Y chromosomes are "horrendously different from each other," says David Page of the Whitehead Institute for Biomedical Research in Cambridge, Massachusetts, who led the work. "If you're marching along the human chromosome 21, you might as well be marching along the chimp chromosome 21. It's like an unbroken piece of glass," says Page. "But the relationship between the human and chimp Y chromosomes has been blown to pieces."

Chapter Eight

Curiosity Kills the Anunnaki

Brainstorming sessions after brainstorming sessions were held.

At length, a strange idea came about.

Enki related the fact to Nimah:

"Perhaps the admixture is not wrong. Maybe we need trace elements, nanolife substances that were indigenous to Earth. They might play a very important role in the functioning and formation of life. We can simulate a life-sustaining indigenous admixture from the clay of Earth. It would serve as a better primordial soup. We should not stick to the conventional thinking of working with test tubes produced by Nibiru factories. Nor should we set up laboratory conditions similar to Nibiru. We must observe biological, chemical, and physical principles interacting here on Earth."

"Then the LORD God formed man of dust from the ground" (Genesis 2:7)

Chapter Nine

FOXP2 Language Gene

Terran Experiment No. 478,911

Nimah made a vessel using Africa's clay. This she made into a primordial soup. In the clay vessel, the fertilized Homo erectus's ova, Anunnaki gene splices, and mitochondria extracts were carefully adjusted to the right proportions. Nimah then inserted the fertilized egg back into the female Homo erectus.

There was conception. A newborn at the awaited time was forthcoming.

It was a male. It was the image of perfection.

The leaders, Ningishzidda, Ninmah, and Enki were overjoyed. Ningishzidda and Enki were backslapping the baby to make him cry, thereby setting the precedent for all human doctors to follow.

The child was growing faster than on their home planet Nibiru. His limbs were suited for a worker. However, he knew not speech. He could only make grunts and snorts.

He had the chromosomal disorder FOXP2 and was unable to select and produce the fine movements with the tongue and lips that were necessary to speak clearly.

The FOXP2 language gene was further investigated and then spliced onto the hybrid's chromosome.

Footnote

In humans, mutations of FOXP2 cause a severe speech and language disorder (Lai CSL, Fisher SE, Hurst JA, Vargha-Khadem F. Monaco AP (2001). "A forkhead-domain gene is mutated in a severe speech and language disorder." *Nature* 413 (6855): 519-23

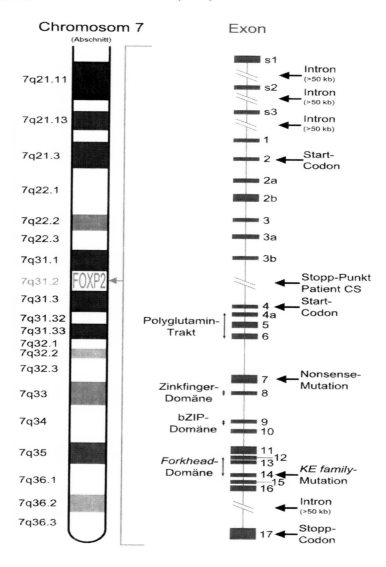

Chapter Ten

Frankenstein in Anunnaki Womb

Enki's kundalini awakening
Enki reconsidered everything.
Reconsidered everything.
Reconsidered everything . . .

From large, structural, chromosomal changes to single nucleotide polymorphisms (SNPs), Enki pondered. Using the latest Anunnaki biogenetic technologies for analysis of genetic variation and genomic profiling, such as array comparative genomic hybridization, Enki realized that there was one single procedure that was always done. That stood in the way of making the perfect being, with all the attributes that they were determined to achieve.

Into the womb of the Earth female, the fertilized ova was always inserted.

"Eureka! This may be the only mistake. We need an Anunnaki gestational mother. An Anunnaki womb!"

Nimah said to Enki, "Who would carry a Frankenstein in her womb?"

"Let me persuade Ninki, my wife," Enki said.

"No, I would dedicate my life in the name of science and face the rewards and endangerment alone!" emphatically said Nimah.

This was the first recorded instance in which a scientist set the example of using herself as the guinea pig of her experiment.

The fertilized egg was inserted into the womb of Nimah.

There was conception.

Would the conception be nine months of Nibiru or Earth time? It was quicker than on Nibiru and shorter than on Earth.

He was born perfect.

The newborn uttered the proper sound upon Enki's slapping of its hind parts.

It was the image of perfection. It was the image of the Anunnaki.

"And God said, "Let us make man in our image, after our likeness." (Genesis 1:26)

Chapter Eleven

Human Prototype Adamu 001

Nimah victoriously shouted, "We've done it!"

A complete physical and medical examination was taken of human prototype Adamu 001. It revealed the following:

Medical Report Checklist of Human Prototype Adamu 001

- His ear was rightly positioned for hearing.
- His eyes were not clogged.
- His lower limbs in time will be muscular.
- His hands in time can hold tools.
- He was not shaggy.
- His hair was dark black.
- His skin was as smooth as that of the Anunnaki.
- His male hood was surrounded by a foreskin, unlike that of an Anunnaki male. And this started the tradition of circumcision in many cultures.
- Adamu was truly created in the *image* of the Anunnaki!

Chapter Twelve

Mass Production of Adamu

Female healers, nurses, doctors, and geneticists were taken to Adamu's incubator.

Volunteers were asked to be gestation mothers. The seven heroines who stepped up to the task were remembered as Ninimma, Shuzianna, Ninmada, Ninbara, Ninmug, Musardu, and Ningunna.

In the male part of Adamu, Nimah made an incision. A drop of blood to let out. She squeezed the male part for blood, one drop of blood in each vessel's admixture.

After in situ hybridization in the wombs of the birth giving heroines, the fertilized ova was inserted.

At the allocated time, seven male earthlings were born. All were healthy.

Seven primitive workers had been created.

The Anunnaki hoped to create the whole race of primitive workers, eventually.

Footnote

The creation story is the same the world over, even though individuals created their own religions and beliefs of their origin. Local species, i.e., gorilla, chimp, orangutan, Japanese monkeys, and others were sampled and their intelligence and physiology genetically placed on the fast track via hybridization with the visiting alien species (i.e., "Let us make them in our own image"). (Genesis 1:26) The subject species would account for racial differences we see today: African negro gorilla, chimpanzee, white Caucasoid orangutan, proboscis monkey, oriental, Japanese monkey. They would have most likely been placed in a quarantine area (Garden of Eden) and other "gardens" mentioned in the world's religions.

Chapter Thirteen

Wanted: Gestation Surrogate Mothers!

Africa News Network Infomercial (*March 449,990 B.C.*)

"There is no greater gift than the gift of life. This is a simple and profound truth.

"Very few women are generous, caring, and selfless enough to become a surrogate. Will you be one to contribute to the worthy cause of the Anunnaki? Please review some basic qualifications and submit a brief application to the Anunnaki Center for Terran Genetics, Garden of Eden. You will be contacted by an experienced female healer who will explain the process to you and answer any questions."

The Surrogate Mothers Program had proven to be an excruciating task on the Anunnaki females. It was also a huge drain on resources and manpower. Ti-Amat, the female companion for Adamu, had to be fashioned so that Adamu and Ti-Amat could reproduce by themselves.

Ninki, Enki's wife, was fascinated to create Ti-Amat. After being given her doctor's release and briefed on the dangers, she stood up to the challenge.

Ti-Amat, the first female earthling, was born.

Chapter Fourteen

Nature's Curse for Hybrid Animals

Adamus and Ti-Amat were wonders of wonders to behold in The Terran Wild Life and Natural Reserve Center in the Garden of Eden.

When the earthling grew up, there were sexual activities but no conception. This fact had the Anunnaki leaders worried a lot.

Ningishzidda ordered his subordinates to place hidden cameras in hidden tents by the trees around the cages where the earthlings lived.

The earthlings' behavior patterns were observed and analyzed.

Mating there was. Conceiving there was not. Birth giving there was not.

Like hybrid animals, ligers, mules, and beefalos, Adamu and Ti-Amat could not procreate. A natural curse had been placed.

Disappointment engulfed the leaders.

They decided to reexamine Adamus and Ti-Amat's chromosomes.

Chapter Fifteen

Genetic Experiment That Goes Horribly Right

The Anunnaki Terran Operations referendum was thrown at the Terran gene splicing quarter. Every healer, geneticist, doctor, and scientist on Earth was invited. There was a consensus that in the Human Genome Project the Anunnaki had been too deeply involved.

In the human genetic experiment that went horribly right, they were giving more and more of their DNA essences to the humans in the hopes of creating a being who possessed the specific characteristics that they desired, that of a primitive intelligent worker to work in the mines. It was originally estimated that a drop of their essence would suffice. Now, there seemed to be no end to the addition of Anunnaki genes . . .

A great effort was carried out to have the sequence of Adamu and Ti-Amat genomes examined, compared, and contrasted with male and female Anunnaki genomes.

It was found out that after separating the double helix there were only twenty-two pairs of chromosomes. There were no sex chromosomes present. And that was nature's curse for hybrid animals.

Footnote

In light of the knowledge explosion and our knowledge of the human brain, the theory that humankind evolved from lower primates gradually over hundreds of thousands of years of time is just not fitting in with the facts. Humans have an exceptionally big brain relative to their body size. Although humans weigh about 20 percent more than chimpanzees, our closest relative, the human brain weighs 250 percent more. How such a massive morphological change occurred over a relatively short evolutionary time has long puzzled biologists.

Evidence That Human Brain Evolution Was a Special Event

Genes that control the size and complexity of the brain have undergone much more rapid evolution in humans than in nonhuman primates or other mammals, according to a new study by Howard Hughes Medical Institute researchers.

The accelerated evolution of these genes in the human lineage was apparently driven by strong selection. In the ancestors of humans, having bigger and more complex brains appears to have carried a particularly large advantage, much more so than for other mammals. These traits allowed individuals with "better brains" to leave behind more descendants. As a result, genetic mutations that produced bigger and more complex brains spread in the population very quickly. This led ultimately to a dramatic "speeding up" of evolution in genes controlling brain size and complexity.

"People in many fields, including evolutionary biology, anthropology, and sociology, have long debated whether the evolution of the human brain was a special event," said senior author Bruce Lahn of the Howard Hughes Medical Institute at the University of Chicago. "I believe that our study settles this question by showing that it was."

David Haussler, director of the Center for Biomolecular Science and Engineering at the University of California, Santa Cruz, said his team found strong but still circumstantial evidence that a certain gene, called HAR1F, may provide an important answer to the question, "What makes humans brainier than other primates?"

Chapter Sixteen

ANNUNAKI CLASSIFIED DOCUMENT: "Genesis"

A great chunk of Anunnaki DNA was given to Homo sapiens against the Anunnaki Grand Council's Ruling to make the Human Genome Project a success.

After the general sessions, behind closed doors, Ningishzidda, Ninmah, and Enki held a secret discussion. The agenda was put into the classified document—"Genesis"—of the highest level. The document detailed the determination of the leaders to go the whole nine yards in creating the perfect being.

The Anunnaki had come too far, been too involved and too entrenched to pull out. There was no way to go back. They had to make this

Human Genome Project a success, even if it meant a great chunk of their DNA would have to be given.

In the meantime, everything had to be carefully kept under the dark in case of sure opposition from Supreme Commander Enlil and his supporters, who were opposed to the creation of earthlings as primitive workers from the start. A special task force of healers, geneticists, doctors, and scientists who were sworn to secrecy was thus formed and placed under Ningishzidda's management.

When the time came, Ningishzidda had everyone leave the Terran genetic splicing lab, except a handful of assistants and himself.

Ningishzidda had induced a deep sleep and applied anesthesia to Ninmah, Enki, Adamu, and Ti-Amat.

From the bone marrow of Enki, he obtained the material for the Y chromosome. From the bone marrow of Ninmah, he obtained the material for the X chromosome. In the adult germ line stem cells of Adamu and Ti-Amat, his secret team of assistants advanced their state-of-the-art Anunnaki genetic biotechnology. After the operations, the wounds in all the ribs of the patients were closed up.

In the Terran genetic splicing lab, with the state-of-the-art technology, patients went home the same day.

(Genesis 2:21-25: "And the LORD God caused a deep sleep to fall on Adam, and he slept: and he took one of his ribs . . . ")

Enki knew the 223 genes in his genome differed from Homo erectus's.

The human genome contains fewer than thirty thousand genes. It contains 223 genes that do not have any predecessors on the genomic evolutionary tree. These 223 genes involve physiological and cerebral functions peculiar to humans. Enki withheld genes affecting longevity. The Anunnaki did not give us the relatively extreme longevity they possessed because it did not suit their purposes. We were invented as slave workers.

There is an interesting parallel in America. In 1998, DNA testing provided answers, which were hard to refute, to the mystery some historians have alleged for years—that Jefferson, after the death of his wife Martha developed a relationship with a slave on his plantation with whom he fathered several children.

In the foreseeable future, earthlings can stake their rightful claim with the Anunnaki.

Footnote

Dr. Francis Crick concluded extraterrestrial origins in the human genome, in relation to his well renowned DNA research. In his book *Life Itself: Its Origins and Nature* (1981), Crick—a Nobel prize winner and the cofounder of the shape of the DNA molecule—claimed an advanced civilization transported the seeds of life to Earth in a spacecraft.

Evolutionary biologist Doolittle: "The most exciting news from the human genome sequencing project has been the claim by the 'public effort' that between 113 and 223 genes have been transferred from bacteria to humans over the course of vertebrate evolution."

One of Australia's best-known scientists has written that our own DNA may contain hidden 'alien' messages from space. Writing in *New Scientist,* Professor Paul Davies, from the Australian Centre for Astrobiology at Macquarie University in Sydney, believes that a cosmic greeting card could have been left in the so-called 'junk DNA' contained in every human cell.

Human Genome Project coordinators find absolute proof of extraterrestrial contact with 'Earth humans' via DNA evidence.

Professor Chang is only one of many scientists and other researchers who have discovered extraterrestrial origins to humanity.

Professor Chang and his research colleagues show that apparent "extraterrestrial programming" gaps in DNA sequencing precipitated by a hypothesized rush to create human life on Earth presented humankind with illogical growth of mass cells we know as cancer." (John Stokes. *Scientists Find Extraterrestrial Genes In Human DNA: Do civilizations of advanced human beings exist scattered in the Galaxy?* John Stokes, 1-11-7. http://rense.com/general74/d3af.htm)

Chapter Seventeen

Expulsion from the Garden of Eden

Adamu and Ti-Amat were again roaming in the Garden of Eden. They became aware of their sexuality and nakedness. Ti-Amat made aprons from leaves of the wild plants for herself and Adamu.

As Enlil was enjoying his leisure walk in the Garden of Eden, he met the couple. He noticed the aprons on their loins. He was perplexed. And Enki was summarily summoned to explain.

"How do they come to this realization? How do they become so cultured and knowledgeable?

"Now, they must have most of our DNA, and maybe they were also given the gene of longevity! Like us they shall live for a long, long time!"

Ningishzidda and Nimah were also summoned for further clarification.

They responded, "Sex chromosomes were given. The genomic branch for longevity was not given."

In fits of anger, Enlil expelled Adamu and Ti-Amat from the Garden of Eden.

Chapter Eighteen

The Descent of Man

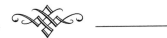

Mass production of prototypes Adamu 001 and Ti-Amat 001 created tribal villages. Sons and daughters were born to Adamu, Ti-Amat, and their prototypes. Before long, a population of primitive slave workers was born. Since they had the ability to follow Anunnaki commands, the astronauts from Nibiru were relieved from duty from their gold mine excavations and replaced by earthlings.

Loads and loads of gold were quickly transported back to Nibiru to make into the finest dust that could be suspended in Nibiru's atmosphere to heal the broken ozone layer. The atmosphere was gradually healing. Planet Nibiru was saved, and the great Anunnaki civilization thrived.

Over time, the children's children of Adamu and Ti-Amat wandered, mixed, and married with other experimental humanoids, thus spoiling Enki's plan of maintaining a quality slave worker race of perfect DNA with perfectly designed attributes.

To counter the bad influence and potential degeneration of population demographics, Enki and his bureau of intellectuals sought to implement the social and hierarchical systems of the Anunnaki civilization to the earthlings.

Footnote

The domesticated silver fox experiment:

Fox pups of ages seven to eight months were classified into four tameness categories. In the tenth generation, 18 percent of foxes were of the tamest category. It the twentieth generation, 25 percent were the tamest, and after forty years, the ratio increased to 70 to 80 percent. It was also established that genetics were responsible for 35 percent of the tameness variation.

The result is that Russian scientists now have a number of domesticated foxes that are fundamentally different in temperament and behavior from their wild forebears. Some important changes in physiology and morphology are now visible, such as mottled or spotted colored fur. Many scientists believe that these changes related to selection for tameness are caused by lower adrenaline production in the new breed, causing physiological changes in very few generations and thus yielding genetic combinations not present in the original species. This indicates that selection for tameness (i.e., low flight distance) produces changes that are also influential on the emergence of other "doglike" traits, such as a raised tail and coming into heat every six months rather than annually. (Wikipedia. "Domesticated silver fox." Accessed October 2013. *http://en.wikipedia.org/wiki/Domesticated_silver_fox*)

Chapter Nineteen

Man's First Society

Fauna and flora were brought from planet Nibiru and given to the earthlings as domesticated animals and plants to harvest.

Knowledge of botany, zoology, geography, mathematics, and theology were imparted in schools. Religions were set up to worship the Anunnaki as gods. Law codes and social reforms were instituted. Kingships were established to help rule over the primitive workers.

The Sumerian language, with precise grammar and rich vocabulary, was simplified from the Anunnaki language.

The first true cities, having ten thousand to fifty thousand inhabitants, sprang up. This sudden civilization, which appeared in full swing in Sumer, Mesopotamia—today's Iraq—over several thousand years ago, became the forbear of all ancient civilizations—Egyptian, Mayan, Indian, Jewish, Greek, Chinese . . .

Footnote

Evidence of the domestication of human beings by Eugen Fischer:

Domestication: the arbitrary influencing of diet and conditions of reproduction.

Morphological similarities between man and domesticated animals.

Domestication led to great variations, especially in terms of size, hair, pigmentation, tails, ears.

The white skin color, as well as the partial loss of pigment, which is responsible for blue and green eyes, represented domesticated albinism. No mammal existing in the wild has a distribution in the pigment of the eye comparable to that of a European. By the same token, among almost all domesticated species, there are individual breeds with eye color identical to a European's.

Domesticated forms, including man, have a pronounced tendency to vary in size. And in domestication size is usually hereditary. A reduction in the facial portion of the skull and a weakening of the teeth are also signs of domestication. Domestication in animals also led to increased appetite, increased and indiscriminate sexuality, and breakdown instincts.

The principal characteristic of domestication is a reduction or even a complete halt to brain development. The opposite has occurred in humans. Another sign of domestication in animals is the early sexual maturity. The same is also not evident in man. (Arnold Gehlen. *Man: His Nature and Place in the World.* 106)

"The male is a domestic animal which, if treated with firmness and kindness, can be trained to do most things."

—Jilly Cooper

Chapter Twenty

Homo Sapiens's Final Genetic Makeover

By this time, the daughters of man were so civilized, beautiful, and sophisticated that the descendants of Anunnaki pursued them. Offsprings with extra intelligence, physical strength, and extraordinary abilities were born. They became ancient heroes and men of renown across all cultures.

"When man began to multiply on the face of the land and daughters were born to them, the sons of God saw that the daughters of man were attractive. And they took as their wives any they chose." (Genesis 6:1-4)

Footnote

Devolution as opposed to evolution:

John C. Sanford, a plant geneticist, has argued for devolution. He has written a book entitled *Genetic Entropy and the Mystery of the Genome* (2005) in which he claims that the genome is deteriorating and therefore could not have evolved in the way specified by the modern evolutionary synthesis. Sanford has published two peer-reviewed papers modeling genetic entropy.

Bizarre examples of genetic engineering:

Glow in the dark cats: Scientists say the ability to engineer animals with fluorescent proteins will enable them to artificially create animals with human genetic diseases.

The enviropig is a pig that's been genetically altered to better digest and process phosphorus.

Pollution fighting plants: poplar trees that can clean up contamination sites by absorbing groundwater pollutants through their roots.

Venomous cabbage: limits pesticide use while still preventing caterpillars from damaging cabbage crops.

Web spinning goats: goats that produce spiders' web protein in their milk.

Fast growing salmon: genetically engineered Atlantic salmon

Banana Vaccines: when people eat a bite of a genetically engineered banana, which is full of virus proteins, their immune systems build up antibodies to fight the disease—just like a traditional vaccine.

Less flatulent cows: cattle that creates 25 percent less methane than the average cow.

Genetically modified trees: trees genetically altered to grow faster, yield better wood, and even detect biological attacks.

Disease fighting eggs: The hens lay eggs that have miR24, a molecule with potential for treating malignant melanoma and arthritis, and human interferon b-1a, an antiviral drug that resembles modern treatments for multiple sclerosis.

Super carbon capturing plants: Carbon contributes to the greenhouse effect and global warming; therefore, researchers hope to create bioenergy crops with large root systems that can capture and store carbon underground. (Mother Nature Network. "12 bizarre examples of genetic engineering." Accessed October 2013. *http://www.mnn. com/green-tech/research-innovations/photos/12-bizarre-examples-of-genetic-engineering/mad-science#)*

Chapter Twenty-One

Anunnaki Xenobiologists' Science Digest

Anunnaki Xenobiologists' Science Digest

July 449,500 B.C. Issue A Study of Hybrid Animal and Homo Sapiens

This whole chapter appeared as the cover page story of a science digest.

A study on the hybrid animals of Earth revealed similarities to Homo sapiens, who were essentially Anunnaki and ape hybrids.

It is extremely rare in the wild, and the majority of the hybrid offspring, ligers (male lion and female tiger) and tigons or tigrons (male tigers and female lions), are bred in captivity.

The lifespan of ligers, as well as other hybrid animals, is shorter than a normal species. The animals seem prone to cancers and other illnesses. It is possible that the mix of genes contributed to the illness.

The size and appearance depends on which subspecies are bred together. The smaller size of the tigress compared to the lion means that some or all of the cubs may be stillborn, or the cubs may be born prematurely (there isn't enough space in the womb for them to develop any further) and may not survive. Premature birth can lead to health problems in those that survive.

Female tigons and ligers have often proved to be fertile and can mate with a lion, tiger, or—in theory—with another species such as a leopard or jaguar. Tigons and ligers have been mated together to produce ti-ligers (tig-ligers). Tigers and tigons have been mated to produce ti-tigons. Ti-ligers and ti-tigons are more tiger like (75 percent tiger). Ti-tigons resemble golden tigers but with less contrast in their markings. It is possible that the mix of genes contributed to the illness . . .

Chapter Twenty-Two

The Anunnaki Confront Changing Demographics: Sumerian Africa News Network, July 449,500 B.C.

Anunnaki Colony's Changing Demographics:

Humans were given the ability to procreate prolifically by Enki, and this had led to an explosion in the human population, which threatened to swamp the Anunnaki, who were never great in numbers.

The dilemma, it seemed, was that these demographic changes were on a trajectory that seemed unlikely to change, which spelled a particular problem for the Anunnaki in years to come

The Anunnaki had at last awakened to its existential crisis. The populous hybrid earthlings were never to Supreme Commander Enlil's liking.

From a global climate forecast, he knew a big flood was coming.

He decided to let humans perish in the impending big global flood.

Chapter Twenty-Three

Global Flood for Humanity's Destruction

Enki was on the human's side, and he had Noah in charge of the prehistoric Svalbard Global Seed Vault Project. To escape Enlil's detection, an ark, called Noah's Ark, was built on the mountains. "The ark came to rest on the mountains of Ararat" (Genesis 8:4).

When the time came, the Anunnaki left the planet in a flying craft, as an enormous surge of water wiped out much of humanity. There was no doubt that an unimaginable catastrophe, or more likely catastrophes, visited upon the Earth between approximately 11,000 and 4,000 BC. The geological and biological evidence is overwhelming in its support of the countless stories and traditions that describe such events. They come from Europe, Scandinavia, Russia, Africa, throughout the American continent, Australia, New Zealand, Asia, China, Japan, and the Middle East.

The seeds of humanity were preserved after the global flood. Enlil was greatly enraged by Enki's double dealing.

As time passed, he regretted his early decision to let humanity perish.

With the ability to quickly multiply, human settlements again reached epic proportions.

Chapter Twenty-Four

Lording over the Humans

In time, colonies grew humongous and spanned many continents on Earth.

The Anunnaki created bloodlines to rule humanity on their behalf, and these are the families still in control of the world to this day.

Kingship was granted to humanity by the Anunnaki, and it was originally known as Anuship after An or Anu, the ruler of the 'gods.'

Eventually, kingships became prevalent all over the globe.

The royal families and aristocracy of Europe, Asia, and the Middle East are obvious examples of this.

Most of the world's organized religions were also established at around this time.

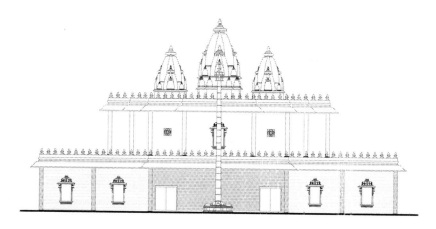

Chapter Twenty-Five

The Quest for Global Domination

The two great rivals for all the major issues facing the Anunnaki colonies eventually became two opposing camps, each controlling a hemisphere.

Enki, though the first born of Anu, was subordinate to Enlil because of the Anunnaki's obsession with genetic purity and laws of succession.

Enlil's mother was the half sister to Anu, and this union passed on the male genes more efficiently than Enki's birth via another mother.

The Anunnaki had many internal conflicts and high-tech wars with each other, as the Enlil and Enki factions fought for control and desired dominance over this planet.

Any group so imbalanced as to covet the complete control of the planet would be warring within itself as different factions sought ultimate control. There were tremendous internal strifes, conflicts, and competitions within the Anunnaki High Command.

Chapter Twenty-Six

Prehistoric World War Three Erupted

In 2024BC, the two opposing camps had a thermonuclear exchange.

The scenario was summarized below.

How smitten is the land, its people delivered to the Evil Wind, its stables abandoned, its sheepfolds emptied.

How smitten are the cities, their people piled up as dead corpses, afflicted by the Evil Wind.

How smitten are the fields, their vegetation withered, touched by the Evil Wind.

How smitten are the rivers. Nothing swims anymore, pure sparkling waters turned into poison.

In its glorious cities, only the wind howls; death is the only smell.

How smitten is the land, home of gods and men!

On that land a calamity fell, one unknown to man.

A calamity that mankind had never before seen, one that could not be withstood.

THE GREATEST STORY NEVER TOLD

On all the lands, from west to east, a disruptive hand of terror was placed. The gods, in their cities, were helpless as men!

(from *Lost Book of Enki* by Zecharia Sitchin, chapter one)

For the next one hundred and fifty years, under Adad's supervision, most of the Nibirans left Earth from Nazca, Peru for Nibiru.

Part 2

The Once and Future Global Civilization

"All things are wearisome, more than
   one can say.
The eye never has enough of seeing, nor
   the ear its fill of hearing.
What has been will be again, what has
   been done will be done again;
there is nothing new under the sun.
Is there anything of which one can say,
   Look! This is something new?
It was here already, long ago; it was here
   before our time.
No one remembers the former
   generations, and even those yet to
   come will not be remembered
by those who follow them."

--King Solomon

Just when we pride ourselves to have come very far to represent the apex of the world's civilizations, having the best forms of democracy, social institutions and the leading edge in technology, we are learning more and more about a prehistoric advanced societies that was planet wide and also highly technical. What we think of as our earliest societies, the Sumerians Egyptians, Babylonians, Greeks, Romans, Indians and Chinese are nothing but the remnants of this earlier prehistoric global civilization. The evidence is everywhere. Avalanche of evidences exploded every now and then in new archaeological finds.

Some 445,000 years ago, Enki, the Most Renowned Genetic Scientist from Planet Nibiru discovered that Erectus-Neanderthal genome could assist acclimatization of Nibiran Genome to Earth ecosystem and environment by the process of genetic hybridization. The successful colonization of Earth as a Mining Operation would be feasible with the creation of a slave species: Homo sapiens. Enki was subsequently mistaken and worshipped as The God by many cultures throughout the world.

Our religions were established by these Anunnakis. Myths, fallacies and wrongful adaptations were passed down from generations after generations. The only evidence on the "Existence" of Jesus Christ would have to be a verifiably contemporary eyewitness account by someone other than an avowed disciple or follower of Jesus (like, say, the chef at the Last Supper, assuming he wasn't Christian). Other than the Bible, Jesus Christ should also be mentioned in many historical writings of his time, or shortly thereafter. There should be voluminous mentioning and cross referencing of Jesus in many of the contemporary books and media of his day. Look at the publicity we have given our celebrities, look at the Media coverage of Prince William and Princess Kate around the globe . . .

The only other non-biblical sources, who wrote their very brief references to Christ at least a century after his purported crucifixion, are those of Josephus and Tacitus. Furthermore, even in the Gospels, there are many inconsistencies. There are many years of unrecorded and unaccounted for life in Jesus's upbringing. The fact is that, until all such hard evidences are produced, Jesus Christ must remain a terrifically inspirational but entirely faith-based, non-historical, fictional character.

The key to our future lies in our understanding of our past..

Chapter One

The Cuneiform Clay Tablets Found in Sumeria

Cuneiform writings had been saved on tens of thousands of clay tablets that have been found in Mesopotamia, in the region of modern Iraq, in the last one hundred and fifty years. Through clay tablets, cylinder seals and stele, the Sumerians have provided us with a graphic and richly detailed version of man's early histories including the story of creation, both of the Earth and of man.

The Sumerian account, through the paraphrasing of scholar Zecharia Sitchin, is really the only account that provides totally a plausible series of events that adequately explains every single puzzle we are faced with. The fact that the Sumerians knew about Uranus, Neptune, and Pluto way before anyone else raises the question: If their story is untrue, then how on Earth were they able to acquire such accurate and detailed knowledge?

Uranus was discovered by William Herschel on March 13th of 1781. This was the first time a planet was discovered using a telescope. Neptune was discovered on Sept 23rd 1846 . . . and only by math prediction due to Uranus' orbit rather than actual observation. Pluto wasn't discovered until 1930.

It appears obvious that both of the latter accounts of Babylonian and Bible creation stories were heavily influenced by the much longer and more detailed earlier Sumerian story and can be easily confirmed, as many parallels stories can be seen in all of them.

From their texts, it was clear that the Sumerians had quite a significant amount of scientific and astronomical knowledge imparted by their creator, the Anunnaki.

Chapter Two

Egyptians Pyramids, the Sphinx and Global Megaliths

The enigma of pyramids and megalithic sites throughout the world has intrigued archeologists, historians, and engineers alike, and none of them have provided logically satisfying answers to the enigma. The very fact that one to two hundred tons of stone blocks had been cut precisely with laser precision and hauled close to five hundred feet above the ground, a feat impossible even with today's technology, is a strong indication of ancient technological achievement. It is something that we cannot dismiss off hand. And these pyramids and megaliths may even be thousands of years older than what has been thought.

- Pyramid Complex. Sedeinga, Sudan
- *One overlooked fact is that Sudan has more pyramids than Egypt.*
- *Ancient kingdoms of Kush and Nubia once rivaled Egypt, Greece and Rome.*
- Pyramid Hill, Lake Baikal, Russia
- Tenerife Pyramids, Canary Islands
- Baalbek, Lebanon

The massive and elegant Roman stonework and columns pale by comparison to the megaliths they were built upon. The temple very visibly incorporates into its foundation stones of some fifteen hundred tons. They are some 68 x 14 x 14 feet! They are the largest worked stones on Earth! It is a mystery how such stones could have been moved into place, even according to our science and engineering knowledge of today. It is also a fact the Romans did not use this type of stonework.

- Mount Etna Pyramids, Sicily, Italy.
- Ligourion Pyramids, Greece
- Pyramids Complex, Xi'an China
- *The world's largest pyramid is rumored to be in Qin Lin County in a "forbidden zone" of China. Estimated at nearly one thousand feet high and made of impounded Earth and clay, holding vast tombs.*
- Almendres Cromlech, Alentejo, Portugal
- Avebury Henge, Avebury, England
- Machu Picchu, Cusco, Peru.
- *There are many odd shaped carved stones. Why would these ancient people build a city at such a high and treacherous altitude, where the valleys below would provide all their needs? Why are there just a few skeletal remains, most of them women, for such a large complex?*
- Callanish Stones, Isle of Lewis, Scotland
- Anta Grande do Zambujeiro, Alentejo, Portugal

- Easter Island, Chile
- *For centuries, scientists have tried to solve the mystery of how the colossal stone statues of Easter Island moved.*
- Carnac Stones, Carnac, France
- Horca del Inca, Copacabana, Bolivia
- Stonehenge, Wiltshire, England
- *Archaeological evidence reveals that pigs were slaughtered at Stonehenge in December and January, suggesting possible celebrations or rituals at the monument around the winter solstice.*
- America's Stonehenge, Salem, New Hampshire, U.S
- Monk Mound, St. Louis, Illinois, U.S.
- Puma Punku, Tiwanaku. Bolivia
- *Careful examination of the stones revealed intricate stonework, as though machine tools or even lasers were used. Puma Punku is located at an altitude of 12,800 feet. No trees grew in that area. The Ceramic Fuente Magna Bowl has Sumerian cuneiform and proto-Sumerian hieroglyphics written on it. There are several stones at Puma Punku that weigh over one hundred tons.*

Many of these large pyramid complexes and megalithic sites have over time been closely examined by engineers, investigators, and experts. What has been noticed is that there are definite signs of tooling and automation. The precision achieved is definitely not handcrafted and requires very high grade engineering equipment.

The precision achieved is definitely not handcrafted and requires very high-grade engineering equipment.

The methodical and meticulous Egyptians kept very careful records about everything they ever did, every king they had, every war they fought, and every structure they built. Why didn't they record how the pyramids were built? Purportedly strange new species of fungi were found in King Tutu's chamber. Again, why?

Chapter Three

Australia's Ancient Egyptian Hieroglyphs

Five thousand years old Egyptian hieroglyphs have been found in New South Wales, Australia.

The two hundred and fifty hieroglyphs are certainly not your average aboriginal animal carvings but something clearly alien in the Australian bush setting. They tell the tale of early Egyptian explorers, injured and stranded, in ancient Australia. The discovery centers around a most unusual set of rock carvings found in the

National Park Forest of Hunter Valley, one hundred kilometers north of Sydney.

The hieroglyphs were extremely ancient, in the archaic style of the early dynasties.

Egyptologist Ray Johnson claimed that these glyphs stemmed from the third dynasty of Egypt and chronicle a tragic saga of ancient explorers shipwrecked in a strange and hostile land.

Having spent many years learning and living with the Bundjalung and Gumilaroi people of northern New South Wales, authors Steven Strong and Evan Strong, after consultation and investigation with elders of Australia, believed they had rediscovered a hidden history on the origins of humanity in Australia and the world. Their books try to prove through scientific facts that which the elders of Australia insist is true.

Their claim is supported by genes, mtDNA, and blood evidence. There are also many experts who concur with them that aboriginal people set sail from Australia, not to, fifty thousand years ago. They claim that aboriginal people sailed to and settled in America over forty thousand years ago and visited many other places including Egypt, Japan, Africa, India, etc. They were the first Homo sapiens who evolved before the sapiens of Africa and who gave the world art, axes, religion, marine technology, culture, cooperative living, language, and surgery.

According to the Strongs, "The debate over whether they were the first people in America is virtually a closed case. Hundreds of bones and skulls have been discovered that are undeniably of 'Australian Aboriginal' origin."

Professor Clive Gamble explains, "We have to construct a completely new map of the world, and how it was peopled."

Chapter Four

200,000 B.C.E Metropolis Found in South Africa

About one hundred and fifty miles inland, west of the port of Maputo, South Africa is the remains of a huge metropolis that measures fifteen hundred square miles within an even larger community of ten thousand square miles. It was estimated to be constructed around 200,000 BCE!

Michael Tellinger tells us, "The photographs, artifacts, and evidence we have accumulated points unquestionably to a lost and never before seen civilization that predates all others—not by just a few hundred years, or a few thousand years . . . but many thousands of years. These discoveries are so staggering that they will not be easily digested by the mainstream historical and archaeological fraternity, as we have already experienced. It will require a complete paradigm shift in how we view our human history."

"The thousands of ancient gold mines discovered over the past five hundred years point to a vanished civilization that lived and dug for gold in this part of the world for thousands of years," says Tellinger. "And if this is in fact the cradle of humankind, we may be looking at the activities of the oldest civilization on Earth. We find roads—some extending a hundred miles—that connected the community and terraced agriculture, closely resembling those found in the Inca settlements in Peru."

The individual ruins mostly consist of stone circles. Most have been buried in the sand and are only observable by satellite or aircraft. Some have been exposed when the changing climate has blown the sand away, revealing the walls and foundations.

Chapter Five

Gobekli Tepe, Turkey

Located about five hundred miles away from Istanbul is Sanliurfa, Turkey, where Gobekli Tepe, the oldest human-made place of worship was discovered. The massive sequence of stratification layers suggests several millennia of activity, perhaps reaching back to the Mesolithic.

- The lowest layer contains monolithic pillars linked by coarsely built walls to form circular or oval structures.
- The middle layer has revealed several adjacent rectangular rooms with floors of polished lime.
- The uppermost layer consists of sediment deposited as the result of agricultural activity.

The monoliths are drawn with pictures of lions, bulls, boars, foxes, gazelles, asses, snakes and other reptiles, insects, arachnids, and birds, particularly vultures and water fowl. The T-shape pillars have carved arms to represent anthropomorphic gods.

The German archeological team has taken fifteen years to uncover about five to 7 percent of a gigantic civilization and proven that Gobekli Tepe is almost twelve thousand years old, much older than Egypt, Greek, Roman, Chinese, Indian, and Hebrew history. No other site is this advanced and is this old.

Graham Hancock says that we have this gigantic site with huge circular megalithic structure, that stands there as a mystery, asking us to go figure how was this done, what's the background to this. We have no idea, who made them. They just come out of the dark of the last ice age, about which we know nothing, and enter this stage of history already fully formed. And to his mind this is indicative of a major forgotten episode of human history.

Since the site of Gobekli Tepe is more than twelve thousand years old, it has extended the first human civilization by more then seven-thousand years from Sumer which predates all ancient civilizations still in existence.

Chapter Six

Armenia's Carahunge Stone Circle Complex

A new study into Armenia's Carahunge stone circle complex has shown that it is one of the oldest known megalithic sites, dated around 5500 BC. According to Russian prehistorian professor Paris Herouni, "Carahunge was created as an astronomical observatory marking the movement not only of the sun and moon, but also the stars. Carahunge's principal stellar alignment is towards Deneb, the bright star in the constellation of Cygnus the swan."

According to Andrew Collins, "Herouni ran the angle of the stone through various astronomical programs and found that it was aligned to Deneb at a date of around 5,500 BC, suggesting that this was the time frame in which Carahunge was in use by an advanced society of astronomer priests. It was this alignment that provided the key to finally dating the site, which was expected to have been constructed during this distant epoch."

Chapter Seven

The Dogon of Mali, West Africa

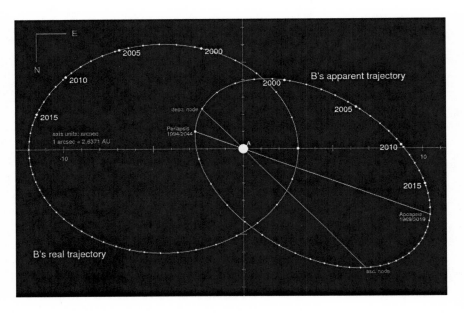

In his book, *The Sirius Mystery,* Robert Temple tells us that in the early twentieth century two French anthropologists named Marcel Griaule and Germain Dieterlen spent a good deal of time living with the Dogon in order to study their ways. In 1930, after they had been living with the tribe for some 15 years, four Dogon Priests decided that it was time to take the Frenchmen into their confidence and invited the men to share in the tribe's most important and sacred tradition. The tale was the secret Dogon creation myths about their sacred star Sirius which located some 8.6 light years from Earth. Sirius is also the brightest star in the night sky. The Dogon told the Anthropologists that Sirius was the home of the Gods who had made them who they are—some eight thousand years ago. They told them that Sirius is the smallest and heaviest thing there is and that it was white in color. They said that it had a companion star, invisible to the human eye but that it moves around Sirius in an elliptical orbit that took fifty years. They said Sirius was incredibly heavy and that it rotated on its axis and they further describe it as having a circle of reddish rays around it that is 'like a spot spreading but staying still.' Dogon oral traditions also quite adamantly state that they have known for thousands of years that Jupiter had moons and Saturn had rings around it.

What they discovered was that the Dogon had in fact, accurately described the three principal properties of a white dwarf star: small, heavy and white and had also stated that Sirius is a binary star, both of which we now know the Sirius system to be. They are also absolutely correct in their knowledge of its companions' rotation as Sirius-B orbits Sirius-A every 49.9 to 50.0 years.

The Dogon people also use an extremely unorthodox calendar that is based on a fifty-year cycle. This cycle is uniquely unusual because it does not follow any cycles coinciding with any movements of our Earth, moon or sun but instead is based wholly on the rotational movements of Sirius B. In fact the entire Dogon Culture is based around the fifty-year cycle of Sirius B. The Dogon People of pre 1930 had no telescopes or real written language. How is it they were able to accurately describe things we still only possessed a very limited knowledge of? Where did they get their information? The Dogon repeatedly say that they were taught these things many, many years ago by their Gods who visited them from their home planet that orbits Sirius B. The Dogon also describe them as being amphibious creatures.

There happens to be too many uncanny concurrences between what the Dogons saw then and what modern scientists see now, and ones that cannot be shrugged off as being mere coincidences.

Part 3

Truths and Misconceptions

Many years have passed since the 2008 financial crisis struck. And only now do Wall Street executives seem to be gaining awareness that many potential clients have lost trust in them and the stock market, and have been voting with their feet. When the stock market was plunging in 2008, some brokers and advisers hid instead of taking calls from clients.

Should we place absolute faith in our financial, scientific and political leaders?

The conclusion is simple: Distrust the experts.

Why? Because you don't know their incentives, and they can make the models say whatever is politically useful to them. This is a manipulation of the public's trust of mathematics, but it is the norm rather than the exception. And modelers rarely if ever consider the feedback loop and the ramifications of their findings or theories on our culture. The truth is somewhat harder to understand, a lot less palatable, and much more important.

This raises a larger question: how can the public possibly sort through all the noise? Whose job is it to push back against rubbish disguised as authoritative scientific theory? When should we stop trusing the icons?

In Sep 2011, the European Centre for Nuclear Research (CERN) showed that the accepted speed of light could be broken by sixty nanoseconds, or more precisely that neutrinos arrived sixty nanoseconds too early than if they were travelling at the speed of light, breaking the speed of light itself by 0.00248%. Two experiments reached the same conclusion. This is a finding that violated Einstein's venerable theory of special relativity that states that nothing in the universe can travel faster than the speed of light in a vacuum.

The journal Science reported that a faulty connection between a GPS and a computer likely caused the anomaly, thus leaving Einstein's theory intact. Scientific findings are not set in stone, and even at the time, CERN doubted its own findings. They agreed that the finding could be an error, saying that even the tiniest of shifts in the Earth between the two measuring points could have changed the results.

General Relativity predicts a singularity at the center of a black hole - most physicists believe this is wrong but no one can prove that yet. A unification of quantum mechanics and General relativity may resolve this.

General Relativity may be wrong in certain circumstances. Rotations of stars in galaxies don't conform to General Relativity. Scientists have 'invented' dark matter to explain this, but since no one knows what dark matter is, it may be that General relativity is wrong under certain conditions. There have been various attempts to add vector and scalar components to Einstein's equations, but so far, they haven't been consistent with observations.

With man's ever-increasing knowledge of science and technology, ancient, half-forgotten legends seemingly have no place. Truths and ideas unable to be examined using orthodox methodology are automatically scorned and ridiculed.

But what happens to the big picture when what we see and what we think we see in the small picture with our limited minds and observations are two very different things?

What are some of our controversial truths and misconceptions?

Chapter One

Mankind Deceived Not Only in Times of Galileo Galilei. The Same Deception Continues . . . Even Today.

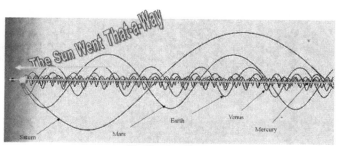

The Planetary Vortex
Movements of the inner planets viewed over a period of one Saturn cycle of 29.36 years.

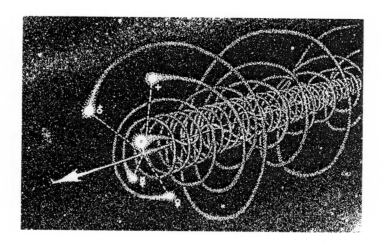

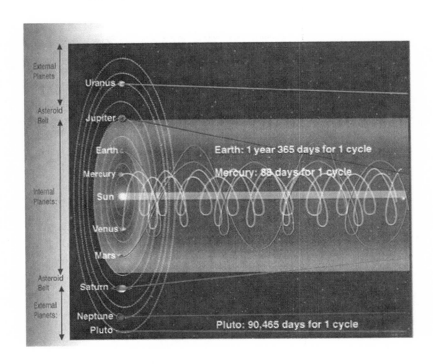

INFO you already know:

- Earth only takes twenty-three hours, fifty-six minutes and 4.09 seconds to turn once on its axis. Astronomers call this a sidereal day.

- Seen from above the north pole, the Earth rotates in a counterclockwise direction. This is why regions in the east see the sun before regions in the west.

- The Earth isn't a perfect sphere. Instead it's an oblate spheroid—a squished ball—because the Earth's bulge points on the equator are a few kilometers more distant from the center of the Earth than the poles.

- The Earth's axis is tilted about 23.5 degrees out of vertical, and that's why we have our seasons. On the June twenty-first northern solstice, the Earth's north pole tilts most toward the sun for the year. Then, six months later, at the December twenty-second southern solstice, it's the Earth's south pole that most tilts toward the sun.

- The Earth revolves around the sun in a nearly circular orbit rather than elliptical because the minimum distance across Earth's orbit is about 98.6% of the maximum distance. That's nearly circular and not elliptical.

- The Earth revolves around the sun once in 365 days, five hours, forty-eight minutes, and forty-six seconds

- The planets move around the sun in elliptical orbits with the sun at one focus. The point in the orbit at which the planet is closest to the sun is called perihelion, and the point at which it is farthest is called aphelion.

Many of us have been taught about how the solar system works by viewing a physical model that has the sun in the middle with the planets going around and around in a simple circular orbit without properly accounting for the motion of the sun (aprox. four hundred and fifty thousand miles per hour). The old model might make one picture be back where you started after a year of time has past, when in fact, you are over eleven BILLION miles from where you were a year ago! This is because the sun revolves around the galactic center.

Earth revolves around the sun, true or false?

Answer:

Believe it or not, there is no empirical evidence that Earth actually orbits the sun! The theory that Earth revolves around the sun leads to the most accurate explanation of the motions we see of all objects in the sky, and it predicts observations that are found when we look for them. Bu it's still "only a theory." If you provide an argument that logically undermines this theory, or another theory that explains what we see in the sky more accurately than this one does, then this theory will be discarded or modified to agree with yours.

This theory has been accepted for about five hundred years. It can never be "proven," but it can be disproved in a minute if you bring in the right kind of new information. That's how science works.

According to Nassim Haramein:

Earth Is Not Orbiting the Sun as We Are Taught

Because of the sun's motion, the planets make those corkscrew helical motions through the galaxy, and therefore heliocentrism is wrong.

In other words, Earth doesn't orbit the sun while the sun stays fixed in space. The Earth and the planets go around the sun, at the same time when the sun is also moving around the galaxy, and the whole cluster moves around the galaxy as a unit. That means sometimes the planets are ahead of the sun and sometimes behind it along that galactic orbit. It seems that Earth returns to the same starting point after 365 days, when in fact, you are over eleven BILLION miles from where you were a year ago!

It is also true to say that if we model the motion of Earth through space over a large number of years, it would be a helical pattern. Our sun is not the center of the galaxy. It revolves around the galactic center. *Therefore, the planets' revolution is spiral rather than circular.*

The analysis of this motion shows that the planets move independently of the sun. Trajectory of motion of the planets is more complicated than the circle and not fit into the theory of revolution of something around something. The nature and trajectory of the **HELICAL REVOLUTION** indicate the existence of invisible energy, which causes the planets to revolve spirally. Partially, this energy is known to us as atomic energy. The motion of electrons is on the same principle. Planets in the solar system revolve under the action of invisible energy. This energy is revolving in a spiral, carrying away for itself material objects.

Chapter Two

Mt. Chimborazo in Ecuador Is Higher Than Mt. Everest by 2,150 Meters / 7,054 Feet

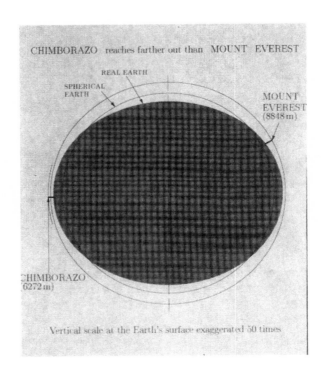

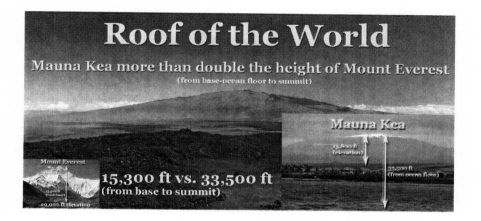

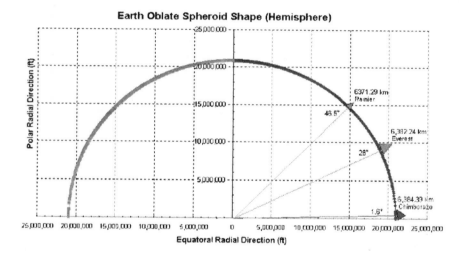

The January 2002 issue of National Geographic magazine printed an interesting confirmation of the validity of Billy Meier's claim back in the 1970s, "Mt Everest is not the highest mountain on Earth." (*Nexus New Times*, July-August 2002, page 61, www.nexusmagazine. com) Meier, in his writings, stated that Mt. Chimborazo in Ecuador was higher than Mt. Everest by 2,150 meters / 7,054 feet because the Earth is not perfectly round but, rather, bulges in the middle. Thus,

Base-to-peak heights of Mount Everest and Mount McKinley (Denali)

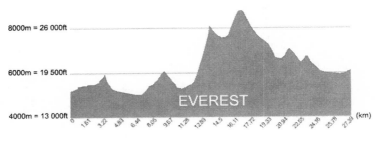

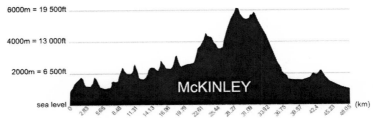

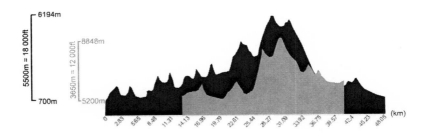

measuring mountains from sea level is not an accurate way of assessing the true height of a mountain.

National Geographic states that scientists have now determined that the Earth bulges around the middle because of the spinning action of the Earth's rotation, and, thus, when measured from the center of the planet, Mt. Chimborazo is actually higher than Mt. Everest by 2,200 meters. (*Nexus New Times*, July-August 2002, page 61, www. nexusmagazine.com) Measured from sea level, Mt. Everest is 2,540 meters higher than Mt. Chimborazo. The news brief states that when measured from the center of the Earth, Mt. Chimborazo is 6,384,450 meters high and Mt. Everest is 6,382,250 meters high.

Chapter Three

Sizes of China, India, Mexico and Africa Are Grossly Wrong on Google World Map Based on Mercator Projection

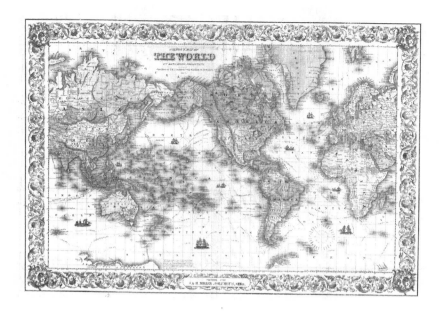

"Each model, like each map, can only show part of the truth. And world maps, make that absolutely clear. To understand "the truth" we need many models, many maps, and many points-of-view." (Dr. George F. Simons, *EuroDiversity: Managing Cultural Differences*)

According to Lloyd Pye, author of "Everything You Know is Wrong":

- Distortions exist to create a two-dimensional map from a three-dimensional globe.
- Google world map, according to Mercator Projection, has a grossly European centrist viewpoint.
- 18.9 million square miles of the **Northern Hemisphere** north of the equator is shown much larger than the 38.6 million square miles of the **Southern Hemisphere** south of the equator.
- 3.93 million square miles of **Europe** is shown much larger than the 6.9 million square miles of **South America**.
- 0.8 million square miles of **Greenland** is shown much larger than the 3.7 million square miles of **China**, which should be about four times the size of **Greenland**.
- 0.4 million square miles of **Scandinavia** is shown dwarfing 1.3 million square miles of **India**, which should be three times larger than **Scandinavia**.
- 0.6 million square miles of **Alaska** are drawn much larger than 0.7 square miles of **Mexico**.
- 8.6 million square miles of the former **USSR** is greatly inflated in contrast to the greatly diminished 11.6 million square miles of **Africa**.
- *According to Dr. George F. Simons, author of EuroDiversity (Managing Cultural Differences):*
- **Indonesia**'s total land area is 741,052 sq mi. It is hard to see from the map that it is about five times that of **Japan**'s

total land area of 145,000 sq. miles. In fact, **Japan** is smaller than the **State of Montana, United States**, which is 147,000 sq. miles.

- Many of the traditional maps like the Mercator exaggerated the goodies for their primary consumers, the developed nations of the Northern Hemisphere, and gave us a Euro/ North American centric world, exactly the problem we are trying to deal with as we train our people to "go global" and come to grips with other folks in our extended human enterprises.

- The distance between many maps and the reality they aim to represent may have resulted historically from a combination of ignorance, self-interest, limited perspectives, and in some cases even deliberate deception. Looking at ancient maps, we often notice that familiar territory is highly detailed and relatively accurate while faraway places with strange sounding names were often quite lumpy and indistinct, even as more remote regions earned the name terra incognita—unknown land.

- Maps are in a sense like two-dimensional photographs, always true but at the same time always deceptive, because they represent a moment in time and can only display two dimensions at a time. And even those two dimensions are not typically on a uniform scale or common unit of measure.

- Sizes of countries and distances in our world are of course best represented by a globe, so attempts at creating flat maps inevitably yield distortions. The old Mercator was designed for navigation and was perfect for directions (lines of constant compass bearing). It also shows shapes somewhat accurately. However, the Mercator distorted the sizes of the land masses terribly.

- "What does it mean if you are not represented on the map?"

McArthur's Universal Corrective Map of the World / South Up Map

Who says north has to be on top?

Stuart McArthur of Melbourne, Australia produced the world's first "modern" south up map and launched it on Australia Day in 1979. Not only does it have south at the top, but it is also "rotated" so that China, Indonesia, and Australia are in the center rather than Europe and West Africa.

Chapter Four

Sun Worshipping Cultures Do NOT Worship Our Sun

The concept of sun worship is one nearly as old as mankind itself. We have been taught that.

- We have Sunday in honor of the sun god.
- The Egyptian peoples honored Ra, the sun god.
- The Greeks honored Helios.
- Native American tribes perform a sun dance each year.
- Ancient Persian societies celebrated the rising of the sun each day.
- Chinese and Korean used the Sun as the ultimate symbol in their philosophy, the concept of yin-yang (light and dark)

However, according to Wayne Herschel, author and symbologist who deciphers Ra Sun symbols in ancient star maps and metal book codices in Jordan: It is not our sun, Solaris, that these ancient cultures worshipped. Historians were mistaken. The ancients revered a sun.

For our ancestors the summer solstice and winter solstice were a time of great spiritual import. However, it was not our sun at all. It was another sun-like star system. Herschel also further states that the gods of the ancients originated from other star systems. And they left star maps for our ancestors, starting in Egypt.

Precession of the equinoxes is a slow westward shift of the equinoxes along the plane of the ecliptic, resulting from the precession of the

Earth's axis of rotation and causing the equinoxes to occur earlier each sidereal year. The precession of the equinoxes occurs at a rate of 50.27 seconds of arc a year; a complete precession requires 25,800 years. Wayne Herschel discovered thirty-eight ancient alien star maps around the world and inferred that alien beings from other star systems taught our ancestors knowledge of mathematics, physics, and astronomy. Ancient humans simply did not have the technology and life-span to observe and document the precession of the equinoxes. It is one of the important legacies and pieces of knowledge that came from alien star beings.

Chapter Five

"X-Men" Superhumans Could Become a Reality in Thirty Years Say MoD Experts

The following is an excerpt from Daily Mail Online by James Rush.

Published on 25 February 2013.

Advancements in gene technology could help humans gain mutant powers such as the likes of Wolverine, Cyclops, and Storm in the popular comic book and movie series, it has been reported.

The MoD's Development Concepts and Doctrine Centre warn however that "genetic inequality" could result from advancements in biology being unequally shared across society.

The center met last summer for a two-day summit, featuring experts from government, industry, and universities. The details have been released following a freedom of information request by *The Sun*. It was reported during the summit, held to predict what would happen in the future, "Advancements in gene technology could lead to a class of genetically superior humans by 2045."

"Human augmentation is likely to increase over the next thirty years." "Discussions highlighted that it is possible that advances in biology, unequally shared across society, could generate genetic inequality.

The X-Men are a team of mutant superheroes created by writer Stan Lee and artist Jack Kirby that first appeared in Marvel Comics in 1963. The mutants use their powers for the benefit of humanity, despite an ever growing anti-mutant sentiment among mankind. The comics were turned into a highly successful film series, featuring Hugh Jackman as Wolverine, Halle Berry as Storm, Ian McKellan as Magneto and Patrick Stewart as Professor X. Halle Berry, playing the character Storm in the popular film series, is able to use her powers to manipulate the weather. Professor X, played by Patrick Stewart in the films, is known as the leader and founder of the X-Men and is able to read, control, and influence human minds.

Chapter Six

Eisenhower's Greada Treaty with the Aliens in 1954

Back in 1954, under the Eisenhower administration, the federal government decided to circumvent the Constitution of the United States and form a treaty with alien entities. The treaty does exist and there are many witnesses to the meeting between aliens and President Eisenhower. Laura Magdalene Eisenhower, the great granddaughter of Pres. Eisenhower also pointed out much circumstantial evidence. Other evidence has been obtained, which is kept in the Library of Congress, proving the meeting with aliens happened, but documentation of the treaty signed by Ike is still many levels above top secret.

This treaty stated the aliens would not interfere in our affairs and we would not interfere in theirs. We would keep their presence on Earth secret; they would furnish us with advanced technology. They could abduct humans on a limited basis for the purpose of medical examination and monitoring, with the stipulation that the humans would not be harmed, would be returned to their point of abduction, and that the humans have no memory of the event. It was also agreed the alien bases would be constructed underground, beneath Indian reservations in the four corners area of Utah, New Mexico, Arizona, and Colorado. Another was to be constructed in Nevada in the area known as S-4, about seven miles south of Area 51, known as "Dreamland." (Michael E. Salla. "False Flag Event in Syria: The Third force and Exopolitics." Posted in "Archaeology, Exopolitics Activism, World Politics," September 9, 2013. http://exopolitics.org/author/dr-michael-salla/)

Chapter Seven

Nature's Favorite Numbers—
Fibonacci Numbers

The Fibonacci Sequence is the series of numbers:
0, 1, 1, 2, 3, 5, 8, 13, 21, 34 . . .

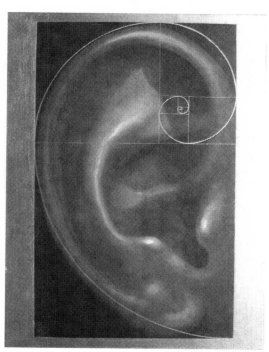

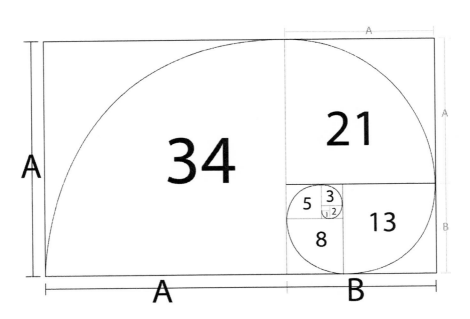

The next number is found by adding up the two numbers before it.

The two is found by adding the two numbers before it (1+1)

Similarly, the three is found by adding the two numbers before it (1+2),

And the five is (2+3), and so on!

The Fibonacci numbers are Nature's numbering system. They appear everywhere in Nature, from the leaf arrangement in plants, to the pattern of the florets of a flower, the bracts of a pinecone, or the scales of a pineapple. The Fibonacci numbers are therefore applicable to the growth of every living thing, including a single cell, a grain of wheat, a hive of bees, and even all of mankind.

Plants do not know about this sequence - they just grow in the most efficient ways. Many plants show the Fibonacci numbers in the arrangement of the leaves around the stem. Some pinecones and fir cones also show the numbers, as do daisies and sunflowers. Sunflowers can contain the number 89, or even 144. Many other plants, such as succulents, also show the numbers. Some coniferous trees show these numbers in the bumps on their trunks. And palm trees show the numbers in the rings on their trunks.

Why do these arrangements occur? In the case of leaf arrangement, or phyllotaxis, some of the cases may be related to maximizing the space for each leaf, or the average amount of light falling on each one. Even a tiny advantage would come to dominate, over many generations. In the case of close-packed leaves in cabbages and succulents the correct arrangement may be crucial for availability of space.

In the seeming randomness of the natural world, we can find many instances of mathematical order involving the Fibonacci numbers themselves and the closely related "Golden" elements.

Fibonacci Petals

3 petals: lily, iris
5 petals: buttercup, wild rose, larkspur, columbine
8 petals: delphiniums
13 petals: ragwort, corn marigold, cineraria
21 petals: aster, black-eyed susan, chicory
34 petals: plantain, pytethrum
55, 89 petals: michelmas daisies, the asteraceae family

The occurrence of Fibonacci Numbers in Nature is interesting but the ratio of consecutive Fibonacci Numbers is important.

Fibonacci in Animals

The shell of the Chambered Nautilus has golden proportions. It is a logarithmic spiral.

The eyes, fins, and tail of the dolphin fall at Golden sections along the body.

A starfish has five arms. (Five is the fifth Fibonacci number)

If a regular pentagon is drawn and diagonals are added, a five-sided star or pentagram is formed. Where the sides of the pentagon are one unit in length, the ratio between the diagonals and the sides is Phi, or the Golden Ratio. This five-point symmetry with Golden proportions is found in starfish.

Humans exhibit Fibonacci characteristics, too. The golden ratio is seen in the proportions of the sections of a finger.

It is also worthwhile to mention that we have 8 fingers in total, five digits on each hand, three bones in each finger, two bones in one thumb, and one thumb on each hand.

The ratio between the forearm and the hand is the Golden Ratio.

The cochlea of the inner ear forms a Golden Spiral.

(Nikhat Parveen, "Fibonacci in Nature," University of Georgia. http://jwilson.coe.uga.edu/emat6680/parveen/fib_nature.htm)

Chapter Eight

Sonic Levitation, Megaliths, and Pyramids, City of Jericho Destroyed by Sonic Vibrations

"In almost every culture where megaliths exist, according to page 432: *Cosmic Key*, a legend also exists that the huge stones were moved by acoustic means—either by the chanted spells of magicians, by song, by striking with a magic wand or rod (to produce acoustic resonance), or by trumpets, gongs, lyres, cymbals or whistles." (Stephen Wagner. *The Ancient Secrets of Levitation*. About.com. http://paranormal.about. com/od/antigravity/a/The-Ancient-Secrets-Of-Levitation.htm)

Did ancient civilizations possess knowledge that has since been lost to science? Were amazing technologies available to the ancient Egyptians that enabled them to construct the pyramids—technologies that have somehow been forgotten?

The ruins of several ancient civilizations—from Stonehenge to the pyramids—show that they used massive stones to construct their monuments. A basic question is why? Why use stone pieces of such enormous size and weight when the same structures could have been constructed with more easily managed smaller blocks, much like we use bricks and cinderblocks today?

Could part of the answer be that these ancients had a method of lifting and moving these massive stones—some weighing several tons—that made the task as easy and manageable as lifting a two-pound brick?

The ancients, some researchers suggest, may have mastered the art of levitation, through sonics or some other obscure method, which allowed them to defy gravity and manipulate massive objects with ease.

Acoustic levitation

"Acoustic levitation uses sound traveling through a fluid—usually a gas—to balance the force of gravity. On Earth, this can cause objects and materials to hover unsupported in the air. In space, it can hold objects steady so they don't move or drift. The process relies on of the properties of sound waves, especially intense sound waves." (Tracy V. Wilson. *How Acoustic Levitation Works*. http://science.howstuffworks. com/acoustic-levitation1.htm)

The Physics of Sound Levitation

A basic acoustic levitator has two main parts: a transducer, which is a vibrating surface that makes sound, and a reflector. Often, the transducer and reflector have concave surfaces to help focus the sound. A sound wave travels away from the transducer and bounces off the reflector. Three basic properties of this traveling, reflecting wave help it to **suspend objects in midair.**

- First, the wave, like all sound, is a longitudinal pressure wave. In a longitudinal wave, movement of the points in the wave is parallel to the direction the wave travels. It's the kind of motion you'd see if you pushed and pulled one end of a stretched Slinky. Most illustrations, though, depict sound as a transverse wave, which is what you would see if you rapidly moved one end of the Slinky up and down. This is simply because transverse waves are easier to visualize than longitudinal waves.

- Second, the wave can bounce off surfaces. It follows the law of reflection, which states that the angle of incidence—the angle at which something strikes a surface—equals the angle of reflection—the angle at which it leaves the surface. In other words, a sound wave bounces off a surface at the same angle at which it hits the surface. A sound wave that hits a surface head on at a ninety-degree angle will reflect straight back off at the same angle. The easiest way to understand wave reflection is to imagine a Slinky that is attached to a surface at one end. If you picked up the free end of the Slinky and moved it rapidly up and then down, a wave would travel the length of the spring. Once it reached the fixed end of the spring, it would reflect off the surface and travel back toward you. The same thing happens if you push and pull one end of the spring, creating a longitudinal wave.

- Finally, when a sound wave reflects off a surface, the interaction between its compressions and rarefactions causes interference. Compressions that meet other compressions amplify one another, and compressions that meet rarefactions balance one another out. Sometimes, the reflection and interference can combine to create a standing wave. Standing waves appear to shift back and forth or vibrate in segments rather than travel from place to place. This illusion of stillness is what gives standing waves their name.

Standing sound waves have defined nodes, or areas of minimum pressure, and antinodes, or areas of maximum pressure. A standing wave's nodes are at the heart of acoustic levitation. Imagine a river with rocks and rapids. The water is calm in some parts of the river, and it is turbulent in others. Floating debris and foam collect in calm portions of the river. In order for a floating object to stay still in a fast-moving part of the river, it would need to be anchored or propelled against the flow of the water. This is essentially what an acoustic levitator does, using sound moving through a gas in place of water.

By placing a reflector the right distance away from a transducer, the acoustic levitator creates a standing wave. When the orientation of the wave is parallel to the pull of gravity, portions of the standing wave have a constant downward pressure and others have a constant upward pressure. The nodes have very little pressure.

In space, where there is little gravity, floating particles collect in the standing wave's nodes, which are calm and still. On Earth, objects collect just below the nodes, where the acoustic radiation pressure, or the amount of pressure that a sound wave can exert on a surface, balances the pull of gravity.

Objects hover in a slightly different area within the sound field depending on the influence of gravity. It takes more than just ordinary sound waves to supply this amount of pressure.

Ed Leedskalnin and Coral Castle

Ed Leedskalnin said science is misled by this false electron/proton principle. He said there are no electrons as perceived. To quote Ed : "The trouble with physicists is they use indirect and ultra indirect methods to come to their conclusions."

Coral Castle is a stone structure created by the Latvian American eccentric Edward Leedskalnin (1887–1951) north of the city of Homestead, Florida in Miami-Dade County. The structure comprises numerous megalithic stones (mostly limestone formed from coral), each weighing several tons. It currently serves as a privately operated tourist attraction. Coral Castle is noted for legends surrounding its creation that claim it was built single-handedly by Leedskalnin using reverse magnetism and/or supernatural abilities to move and carve numerous stones weighing many tons.

Two magnetic currents are the fabric of what is holding together everything in our universe, including the atoms themselves. This is what builds and holds an atom into a structure. The ancients knew this fantastic secret, as did Ed Leedskalnin, and there is a ton of evidence. The message here not to be ignored is Ed Leedskalnin had amazing insight and has given us the true meaning of sacred geometry, and that it is a representation of two magnetic currents. And he left us the visual message before he died.

Biblical city of Jericho B.C. 1400

"An acoustic weapon disorients rioters and afflicts an invading army with nausea. It can create 'ghosts' and arouse animal passions.

"Possibly the earliest account in Western literature of sound itself being used as a weapon can be found in the Bible. As detailed in Joshua 6:5, Joshua leads an attack on the city of Jericho during which he commands his people, outside the walled city, to remain in total silence for seven days. On the seventh day, seven trumpets made from ram's horns give a "long blast," the people shout . . . and the walls of Jericho come crashing down.

"Sound is a waveform, with low infrasonic frequencies having a long wave length (measured in tens of meters), and with high ultrasonic frequencies having a short wave length (measured in millimeters). The frequencies associated with ultrasound are most familiar from their utilisation by the medical profession, chiefly for diagnostic imaging.

"While the ears are designed to detect a limited range of frequencies – the human auditory range is between 20Hz and 20,000Hz (1Hz = 1 cycle per second) – different frequencies can affect the whole body and, at volume, can be felt in almost any part of the body. Even with industrial ear protectors, sound waves are able to enter the head via the nose and mouth that are, in turn, linked to the ears by the structure of the skull. Sounds that are higher in frequency than 20,000Hz – ultrasound – are inaudible to humans, while sounds lower than 20Hz – infrasound – are inaudible but can, on occasion, be felt resonating within the body itself. Exposure of unprotected ears to infrasound can also cause an increase in pressure within the middle ear, disturbing the sense of balance." (Jack Sargeant. "Sonic Weapons." Forteantimes. com. December 2001. http://www.forteantimes.com/features/ articles/256/sonic_weapons.html)

Chapter Nine

The Electric Universe Theory

Today, nothing is more important to the future and credibility of science than liberation from the gravity-driven universe of prior theory. A mistaken supposition has not only prevented intelligent and sincere investigators from seeing what would otherwise be obvious, it has bred indifference to possibilities that could have inspired the sciences for decades.

~David Talbott and Wallace Thornhill, Thunderbolts of the Gods

The Electric Universe

Posted on <u>August 23, 2011</u> by <u>David Talbott</u>

Readers may be surprised to discover that many well-trained skeptics do not support popular ideas in astronomy and the space sciences. Critics doubt that "black holes" actually exist. They suggest that "dark matter," supposedly far more abundant than visible matter, is a mere fiction, hiding the fact that earlier theories no longer work. Theories of galaxy formation, the birth of stars, and the evolution of our planetary system are all raised to doubt by critics who believe that a fateful turn in twentieth century theory set astronomy on a dead-end course.

Enchanted by the role of gravity in the cosmos, astronomers failed to recognize the pervasive role of charged particles and electric currents in space. Electric Universe is a new vantage point, one that acknowledges the contribution of the electric force to the dynamic structure and highest energy events in the universe. As we compare events in space to the behavior of charged particles in the laboratory, the differences between an electric model and the traditional gravity only model should become progressively more clear.

Essential Guide to the Electric Universe

Posted on September 2, 2011 by Bob Johnson - Jim Johnson

The New Picture of Space

Now more than ever, the exploration of our starry universe excites the imagination. Never before has space presented so many pathways for research and discovery.

New observational tools enable us to "see" formerly invisible portions of the electromagnetic spectrum, and the view is spectacular. Telescope images in X-ray, radio, infrared and ultraviolet light reveal exotic structure and intensely energetic events that continually redefine the quest as a whole.

Spectrographic interpretation has grown hand in hand with faster, large memory computers and programs in sophistication and in broad scientific data processing, imaging, and modeling capability.

Standing out amidst an avalanche of new images is the greatest surprise of the space age: evidence for pervasive electric currents and magnetic fields across the universe, all connecting and animating what once appeared as isolated islands in space. The intricate details revealed are not random, but exhibit the unique behavior of charged particles in plasma under the influence of electric currents.

The telltale result is a complex of magnetic fields and associated electromagnetic radiation. We see the effects on and above the surface of the sun, in the solar wind, in plasma structures around planets and moons, in the exquisite structure of nebulas, in the high energy jets of galaxies, and across the unfathomable distances between galaxies.

Thanks to the technology of the twentieth century, astronomers of the twenty-first century will confront an extraordinary possibility. The evidence suggests that intergalactic currents, originating far beyond the boundaries of galaxies themselves, directly affect galactic evolution. The observed fine filaments and electromagnetic radiation in intergalactic and interstellar plasma are the signature of electric currents. Even the power lighting the galaxies' constituent stars may indeed be found in electric currents winding through galactic space.

It was long thought that only gravity could do "work" or act effectively across cosmic distances. But perspectives in astronomy are rapidly changing. Specialists trained in the physics of electricity and magnetism have developed new insights into the forces active in the cosmos. A plausible conclusion emerges. Not gravity alone, but electricity and gravity have shaped and continue to shape the universe we now observe.

The Limits of Gravitational Theory

The Law of Gravity, which relies exclusively on the masses of celestial bodies and the distances between them, works very well for explaining planetary and satellite motions within our solar system. But when astronomers tried to apply it to galaxies and clusters of galaxies, it turns out that nearly 90 percent of the mass necessary to account for the observed motion is missing.

The trouble began in 1933 when astronomer Fritz Zwicky calculated the mass-to-light ratio for eight galaxies in the Coma Cluster of the Coma Berenices ("Berenices's hair") constellation. At the time, it was assumed that the amount of visible light coming from stars should be proportional to their masses (a concept called "visual equilibrium").

As Zwicky was to realize, the apparent rapid velocities of the galaxies, around their common center of mass ("barycenter"), suggested that much more mass than could be seen was required to keep the galaxies from flying out of the cluster.

Zwicky concluded that the missing mass must therefore be invisible or "dark." Other astronomers, such as Sinclair Smith (who performed calculations on the Virgo Cluster in 1936) began to find similar problems. To make matters worse, in the 1970s, radial velocity plots (radius from the center versus stars' speed of rotation) for stars in the Milky Way galaxy revealed that the speeds flatten out rather than trail down, implying that velocity continues to increase with radius, contrary to what Newton's Law of Gravity predicts for, and which is observed in, the Solar System.

In short, astronomers using the gravity model were forced to add a lot more mass to every galaxy than can be detected at any wavelength. They called this extra matter "dark;" its existence can only be inferred from the failure of predictions. To cover for the insufficiency they gave themselves a blank check, a license to place this imagined stuff wherever needed to make the gravitational model work.

Other mathematical conjectures followed. Assumptions about the red shift of objects in space led to the conclusion that the universe is expanding. Then other speculations led to the notion that the expansion is accelerating. Faced with an untenable situation, astronomers postulated a completely new kind of matter, an invisible "something" that repels rather than attracts. Since Einstein equated mass with energy ($E = mc^2$), this new kind of matter was interpreted as being of a form of mass that acts like pure energy—regardless of the fact that if the matter has no mass it can have no energy according to the equation. Astronomers called it "dark energy," assigning to it an ability to overcome the very gravity on which the entire theoretical edifice rested.

Dark energy is thought to be something like an electrical field, with one difference. Electric fields are detectable in two ways: when they accelerate electrons, which emit observable photons as synchrotron and Bremsstrahlung radiation, and by accelerating charged particles as electric currents, which are accompanied by magnetic fields, detected through Faraday rotation of polarized light. Dark energy seems to emit nothing and nothing it purportedly does is revealed through a magnetic field. One suggestion is that some property of empty space is responsible. But empty space, by definition, contains no matter and therefore has no energy. The concept of dark energy is philosophically unsound and is a poignant reminder that the gravity only model never came close to the original expectations for it.

Taking the postulated dark matter and dark energy together, something on the order of twenty-four times as much mass in the form of invisible stuff would have to be added to the visible, detectable mass of the universe. That's to say, in the gravity model all the stars and all the galaxies and all the matter between the stars that we can detect only amount to a minuscule 4 percent of the estimated mass:

Critics often point out that a theory requiring speculative, undetectable stuff on such a scale also stretches credulity to the breaking point. Something very real, perhaps even obvious, is almost certainly missing in the standard gravity model.

Is it possible that the missing component could be something as familiar to the modern world as electricity?

Chapter Ten

Directed Panspermia Proposed by Francis Crick

To overcome the huge hurdles of evolution of life from non-living chemicals on Earth, Crick proposed, in a book called *Life Itself*, that some form of primordial life was shipped to the Earth billions of years ago in spaceships—by supposedly "more evolved" alien beings.

Michael Denton says in his best-selling book *Evolution a Theory in Crisis*, "Nothing illustrates clearly just how intractable a problem the origin of life has become than the fact that world authorities can seriously toy with the idea of Panspermia."

The following is taken from Panspermia-Theory.com, "Recent Support for Panspermia." http://www.panspermia-theory.com/

A meteorite blasted off from the surface Mars about 15 million years ago was found in Antarctica in 1984 by a team of scientists on an annual United States government mission to search for meteors. The meteor was named Allan Hills 84001 (ALH84001). In 1996 ALH84001 was shown to contain structures that may be the remains of terrestrial nanobacteria. The announcement, published in the journal Science by David McKay of NASA, made headlines worldwide and prompted United States President Bill Clinton to make a formal televised announcement marking the event and expressing his commitment to the aggressive plan in place at the time for robotic exploration of Mars. Several tests for organic material have been performed on ALH84001 and amino acids and polycyclic aromatic hydrocarbons (PAH) have been found. However, most experts now agree that these are not a definite indication of life, but may have instead been formed abiotically from organic molecules or are due to contamination from contact with Antarctic ice. The debate is still ongoing, but recent advances in nanobe research has made the find interesting again.

The announcement of the discovery of evidence of life on ALH84001 sparked a surge in support for the theory of Panspermia. People began to speculate about the possibility that life originated on Mars and was transported to Earth on debris ejected after major impacts. On April 29, 2001, at the 46th annual meeting of the International Society for Optical Engineering (SPIE) in San Diego, California, Indian and British researchers headed by Chandra Wickramasinghe presented evidence that the Indian Space Research Organisation had gathered air samples from the stratosphere that contained clumps of living cells. Wickramasinghe called this "unambiguous evidence for the presence of clumps of living cells in air samples from as high as 41 kilometers, well above the local tropopause, above which no air from lower down would normally be transported." A reaction report from NASA Ames doubted that living cells could be found at such high altitudes, but noted that some microbes can remain dormant for millions of years, possibly long enough for an interplanetary voyage within a solar system.

On May 11, 2001, Geologist Bruno d'Argenio and molecular biologist Giuseppe Geraci from the University of Naples announced the finding of extraterrestrial bacteria inside a meteorite estimated to be over 4.5 billion years old. The researchers claimed that the bacteria, wedged inside the crystal structure of minerals, had been resurrected in a culture medium. They asserted that the bacteria had DNA unlike any on Earth and had survived when the meteorite sample was sterilized at high temperature and washed with alcohol. The bacteria were determined to be related to modern day Bacillus subtilis and Bacillus pumilus bacteria, but appear to be a different strain.

On April 21, 2008, renowned British astrophysicist Stephen Hawking spoke about Panspermia during his "Why We Should Go Into Space" lecture for NASA's fiftieth Anniversary lecture series at George Washington University.

In a virtual presentation on Tuesday, April 7, 2009, Stephen Hawking discussed the possibility of building a human base on another planet and gave reasons why alien life might not be contacting the human race, during his conclusion of the Origins Symposium at Arizona State University. Hawking also talked about what humans may find when venturing into space, such as the possibility of alien life through the theory of Panspermia, which says that life in the form of DNA particles can be transmitted through space to habitable places.

Part 4

Be Skeptical Even of One's Own Skepticism

We all know the famous story surrounding Sir Isaac Newton. He was busy discovering the universal law of gravitation, when an apple fell on his head from the branches above him. The fallen apple provided him the inspiration for the laws of gravity. But while the falling apple is a good story, it probably never happened. The story was first published in an essay by Voltaire, long after Newton's death. It was almost certainly an invention.

Centuries-old manuscripts also revealed the hidden pursuits of Sir Isaac Newton when the program "Newton's Dark Secrets" was aired November 15, 2005 at 9 p.m. on PBS. While Isaac Newton was busy discovering the universal law of gravitation, he was also searching out hidden meanings in the Bible and pursuing the covert art of alchemy.

Discoveries were made that showed the unknown side of Newton to the rest of the world. Discoveries that showed Isaac Newton was obsessed with religion and devoted to the occult. Our modern interpretation of Newton is about as far fetch as could possibly be from what Newton himself thought.

Known as the Greatest Scientist of all times who invented calculus; figured out the composition of light; and gave us the laws of gravity and motion, which govern the universe; Sir Isaac Newton was actually pursuing an activity which we now label as a pseudoscience.

The wars for truths are fought relentlessly and ruthlessly over generations. Fear cannot save us. Rage cannot help us. What misfortune or bliss does ignorance hold in store for us in the next hundred years?

There is no sound, no voice, no cry in all the world that can be heard about the truth . . . until someone listens.

Chapter One

Map of Antartica Existed, Three Years Before Its Discovery

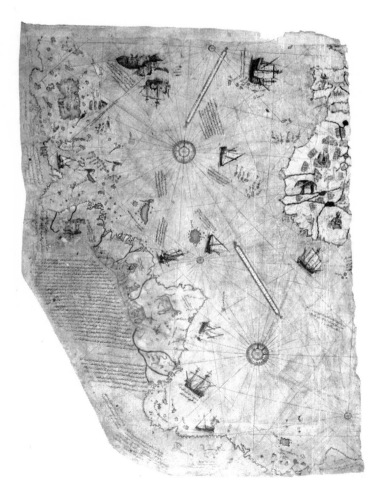

The Piri Reis map shows the western coast of Africa, the eastern coast of South America, and the northern coast of Antarctica. The northern coastline of Antarctica is perfectly detailed. The most puzzling however is not so much how Piri Reis managed to draw such an accurate map of the Antarctic region three hundred years before it was discovered, but that the map shows the coastline under the ice. Geological evidence confirms that the latest date Queen Maud Land could have been charted in an ice-free state is 4000 BC.

The question is: Who mapped the Queen Maud Land of Antarctic six thousand years ago? Which unknown civilization had the technology or the need to do that? It is well-known that the first civilization, developed in the mid-east around year 3000 BC, was soon followed within a thousand years by the Indus valley and the Chinese ones. So, accordingly, none of the known civilizations could have done such a job. Who was here in 4000 BC, able to do things that NOW are possible with modern technologies?

Chapter Two

Pre-Colombian Space Shuttle Models and Ancient Indian Flying Crafts

Pre-Columbian funerary pendants —Gold trinkets of pre-Colombian space shuttle models were found in an area covering Central America and coastal areas of South America, estimated to belong to a period between 500 and 800 CE. However, since they are made from gold, accurate dating is impossible and based essentially on stratigraphy, which may be deceptive. However, we can safely say that these gold trinkets of pre-Colombian space shuttle models are more than one thousand years old.

Clearly, these fascinating objects display aerodynamic features, complete with delta wings, tail stabilizers and elevators, fuselage, and a rudder. They are definitely not stylized figures of fish, birds, frogs, bats, or insects.

In the Vedic literature of India, there are many descriptions of flying machines that are generally called vimanas. Many of the documents contain texts that seem to describe modern aerodynamic principles. Vymaanika-Shaastra talked about metals, electricity, and power sources used for these air crafts. For vimanas pilots, they had special uniforms, weapons that were kept hidden on these air ships, and special flight manuals for references. These fall into two categories:

- Man-made craft that resemble airplanes and fly with the aid of birdlike wings. They are described mainly in medieval, secular Sanskrit works dealing with architecture, automata, military siege engines, and other mechanical contrivances.
- Unstreamlined structures that fly in a mysterious manner and are generally not made by human beings. They are described in ancient works such as the Rg Veda, the Maha-bha-rata, the Rama-yana, and the Pura-nas, and they have many features reminiscent of modern flying machines.

Like many similar literatures and artifacts found in museums, they portrayed the might of ancient aeronautic powers. As more and more ancient artifacts and petroglyphs with pictures of UFOs, wormholes, stargates and alien beings are found throughout the world, the full picture of the jigsaw puzzle soon emerges.

Chapter Three

Sunken Cities of the Caribbean, France, India, and Japan

All around the world, whether it's the Caribbean, France, India, or Japan, we find evidence of a manmade structure that lie under water. One thing to keep in mind is that most of these cities did not sink; they were submerged when the sea level rose. It's really sad that these so called hardheaded "scholars" have their heads in the sand and try to discredit these people who bring to light all of these awesome discoveries. If left to these so called "scholars" we would never have found King Tut's Tomb.

It is not the religious community that fears the undiscovered discoveries that suddenly become discovered. It is the scientific community. It is the scientists who fear being humiliated when found wrong. We know of no Stone Age civilization capable of creating such gigantic ceremonial complexes. If according to the mainstream scholars, the ancient Indian, French, and Japanese could not have created such a structure, then who did? And how did it get there?

Could the ancient underwater cities have been built to serve as bases for ancient aliens? Could these hold the answers that mankind have been seeking for centuries? Evidence that proves we are not alone. These underwater structures and cities seem to just stand out as out of place, in time, and there are questions about them that might throw a new light on the genesis of human history.

I think it's high time the truly knowledgeable archeologists and oceanographers are given a free hand to explore and bring forth as much knowledge of these ancient treasures to be shared with mankind.

Chapter Four

Lessons for Japan and Russia from Ancient Nuclear Reactors

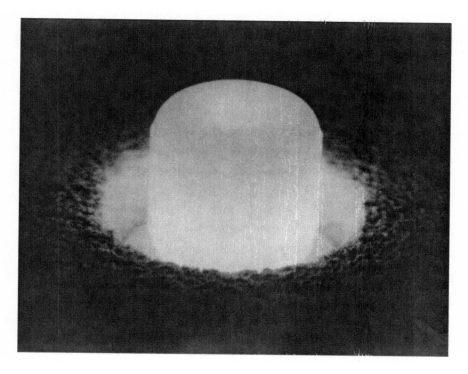

The remnants of nuclear reactors nearly two billion years old were found in 1972 in Oklo in the Gabon Republic in Africa. Surprisingly the uranium concentration in the ore was as low as spent uranium fuel from a nuclear reactor. The finding led scientists to believe that the uranium had already been used for energy production.

This discovery shocked the world and attracted scientists from many countries to go to Oklo for further investigation. The results showed that the uranium mine was an ancient nuclear reactor. The reactor was perfectly preserved, and its layout was very rational. It is estimated that the reactor had been in operation for around five hundred thousand years. Furthermore, nuclear wastes produced in this reactor had not spread all over the surrounding areas. Instead, they were confined within the separate sections. From the perspective of modern nuclear technology, this ancient reactor used very advanced techniques. Oklo by-products are being used today to study the stability of the fundamental constants over cosmological time scales and to develop more effective means for disposing of human manufactured nuclear waste.

Human beings have only made use of nuclear power for a couple of decades. This discovery raised the intriguing possibility that a technologically more advanced civilization existed two billion years ago, and it had advanced knowledge of nuclear fission.

If we neglect relics of prehistoric civilization, there is no way we can broaden the scope of our present knowledge. We will neither know what caused prehistoric civilizations to degenerate, or how they finally came to disappear. Moreover, we should carefully examine whether our current method of scientific development is following the same disastrous road. This is surely a subject worthy of serious consideration.

The long-term preservation of the Gabon natural nuclear reactors is perhaps even more remarkable than the reactors themselves. These nuclear reactors have survived two billion years of geologic time. It is certainly the most valuable lesson from ancient history if you just consider those man-made nuclear disasters like the Fukushima Daiichi nuclear disaster of 2011 and the Chernobyl disaster of 1986.

Ancient Nuclear Explosion

According to British researcher David Davenport, author of *Atomic Destruction in 2000 B.C.*, Milan, Italy, 1979:

An ancient, heavily populated city called Mohenjo Daro in Pakistan was instantly destroyed in 2,000 B.C. by an incredible explosion that could only have been caused by an atomic bomb.

The horrible, mysterious event was recorded in an old Hindu manuscript called the Mahabharata,

- White hot smoke that was a thousand times brighter than the sun rose in infinite brilliance and reduced the city to ashes.
- Water boiled . . . horses and war chariots were burned by the thousands . . .
- The corpses of the fallen were mutilated by the terrible heat so that they no longer looked like human beings . . .
- "It was a terrible sight to see . . . never before have we seen such a ghastly weapon."

What was found at the site of Mohenjo Daro corresponds exactly to Nagasaki.

- Forty-four human skeletons were found there in 1927, just a few years after the city was discovered.
- The city ruins reveal the explosion's epicenter, which measures fifty yards wide.
- At that location everything was crystallized, fused, or melted.
- "Sixty yards from the center the bricks were melted on one side indicating a blast . . . "

Davenport's intriguing theory has met with intense interest in the scientific community.

- How did man in 2000 B.C. have the knowledge of not only producing such a high degree of heat, but also harnessing the power of such high temperatures?
- If Mohanjo Daro was destroyed by a nuclear catastrophe, who designed and manufactured them?
- If not then what was used to produce such heat that vitrified rocks and bricks?
- What could be attributed to the high degree of radioactive traces in the skeletons?
- How did all of them die, in one instant?

Chapter Five

A. 7 Year Old Russian Boy Educated People on Astronomy, Mars Landscape and How to Create High Tech Spacecraft

Sometimes, some children are born with quite fascinating talents, unusual abilities.

http://english.pravda.ru/science/mysteries/05-03-2008/104375-boriska_boy_mars-0/

A boy named Boris Kipriyanovich, or Boriska, lives in the town of Zhirinovsk of Russia's Volgograd region. He was born on January 11, 1996. Since he was four he used to visit a well-known anomalous zone, commonly referred to as Medvedetskaya Gryada – a mountain near the town. It seems that the boy needed to visit the zone regularly to fulfill his needs in energy.

Boriska's parents, nice, educated and hospitable people, are worried about their son's fascinating talents. They do not know how others will perceive Boriska when he grows up. The say that they would be happy to consult an expert to know how to raise their wunderkind.

Being a doctor, his mother could not help but notice that the baby boy could hold his head already in 15 days after his birth. He uttered the first word 'baba' when he was four months old and started to pronounce simple words soon afterwards. At one year and a half he had no difficulties in reading newspaper headlines. At age of two years he started drawing and leaned how to paint six months later. When he turned two, he started going to a local kindergarten. Tutors immediately noticed the unusual boy, his uncommon quickwittedness, language skills and unique memory.

However, his parents witnessed that Boriska acquired knowledge not only from the outer world, but through mysterious channels as well. They saw him reading unknown information from somewhere.

"No one has ever taught him," Boriska's mother said. "Sometimes he would sit in a lotus position and start telling us detailed facts about Mars, planetary systems and other civilizations, which really puzzled us," the woman said.

How may a little boy know such things? Space became the permanent theme of his stories when the boy turned two years old. Once he said that he used to live on Mars himself. He says that the planet is inhabited now too, although it lost its atmosphere after a mammoth catastrophe. The Martians live in underground cities, Boriska says.

The boy also says that, he used to fly to Earth for research purposes when he was a Martian. Moreover, he piloted a spaceship himself. It took place in the time of the Lemurian civilization. He speaks about the fall of Lemuria as if it occurred yesterday. He says that Lemurians died because they ceased to develop themselves spiritually and broke the unity of their planet.

When his mother brought him a book entitled *"Whom We Are Originated From"* by Ernest Muldashev, he got very excited about it. He spent a long time looking through the sketches of Lemurians, pictures of Tibetan pagodas, and then he told his parents of Lemurians and their culture for several hours non-stop. As he was talking, his mother noticed that Lemurians lived 70,000 years ago and they were nine meters tall… "How can you remember all this?" the woman asked her son. "Yes, I remember and nobody has told me that, I saw it," Boriska replied.

In Muldashev's second book *"In Search of the City of Gods"* he looked through pictures for a long time and recollected a lot about pyramids and shrines. Then he claimed that people would not find ancient knowledge under the Great Pyramid of Cheops. The knowledge will be found under another pyramid, which has not been discovered yet. "The human life will change when the Sphinx is opened, it has an opening mechanism somewhere behind the ear, I do not remember exactly," he said.

Boriska is one of so-called indigo children. They start to appear on Earth as a token of the forthcoming grand transformation of the planet.

"No, I have no fear of death, for we live eternally. There was a catastrophe on Mars where I lived. People like us still live there. There was a nuclear war between them. Everything burnt down. Only some of them survived. They built shelters and created new weapons. All materials changed. Martians mostly breathe carbon dioxide. If they flew to our planet now, they would have to spend all the time standing next to pipes and breathing in fumes," Boriska said.

"If you are from Mars, do you need carbon dioxide?"

"If I am in this body, I breathe oxygen. But you know, it causes aging."

Specialists asked the boy why man-made spacecraft often crash as they approach Mars. "Martians transmit special signals to destroy stations containing harmful radiation," Boriska replied.

The boy has deep knowledge of space and its dimensions. He is also aware of the structure of interplanetary UFOs. He talks about that like an expert, draws UFOs on slates and explains the way they work. Here is one of his stories: "It has six layers. The upper layer of solid metal accounts for 25 percent, the second layer of rubber – 30 percent, the third layer of metal – 30 percent, and the last layer with magnetic properties – 4 percent. If we give energy to the magnetic layer, spaceships will be able to fly across the Universe."

Boriska has a lot of difficulties with school. After an interview he was taken to the second grade, but soon they tried to get rid of him. He constantly interrupts teachers and says that they are wrong... now the boy has classes with a private tutor.

Translated by Julia Bulygina

Pravda.ru

B. 12-Year-Old Genius Jacob Barnett Sets Out to Disprove Einstein

By Zachary Roth
Yahoo News

In some ways, Jacob Barnett is just like any other 12-year-old kid. He plays Guitar Hero, shoots hoops with his friends, and has a platonic girlfriend.

But in other ways, he's a little different. Jake, who has an IQ of 170, began solving 5,000-piece jigsaw puzzles at the age of 3, not long after he'd been diagnosed with Asperger's syndrome, a mild form of autism. A few years later, he taught himself calculus, algebra, and geometry in two weeks. By 8, he had left high school, and is currently taking college-level advanced astrophysics classes—while tutoring his older classmates. The 12-year-old taught himself calculus, algebra and geometry in two weeks, and can solve up to 200 numbers of Pi. And he's being recruited for a paid researcher job by Indiana University.

Now, he's at work on a theory that challenges the Big Bang—the prevailing explanation among scientists for how the universe came about. It's not clear how developed it is, but experts say he's asking the right questions.

"The theory that he's working on involves several of the toughest problems in astrophysics and theoretical physics," Scott Tremaine of Princeton University's Institute for Advanced Studies—where Einstein (pictured) himself worked—wrote in an email to Jake's family. "Anyone who solves these will be in line for a Nobel Prize."

It's not clear where Jake got his gifts from. "Whenever I try talking about math with anyone in my family," he told the Indianapolis Star, "they just stare blankly."

But his parents encouraged his interests from the start. Once, they took him to the planetarium at Butler University. "We were in the crowd, just sitting, listening to this guy ask the crowd if anyone knew why the moons going around Mars were potato-shaped and not round," Jake's mother, Kristine Barnett, told the Star. "Jacob raised his hand and said, 'Excuse me, but what are the sizes of the moons around Mars?' "

After the lecturer answered, said Kristine, "Jacob looked at him and said the gravity of the planet ... is so large that (the moon's) gravity would not be able to pull it into a round shape."

"That entire building ... everyone was just looking at him, like, 'Who is this 3-year-old?'"

Chapter Six

Irrefutability of DNA Testing as 99.99% Accurate Shattered by Existences of Natural Human Chimeras

These cases of natural human chimeras challenge the blind faith which the scientific community places on the irrefutability of DNA testing. Forensic science cannot rely on DNA testing as the sole source of evidence, as it has done previously, as the criminal or victim may be a chimera. It follows that current maternity and paternity testing methods will have to be re-evaluated.

Even though there are two or more different sets of DNA in human chimeras, it may or may not be manifested as physical abnormalities. It may appear as phenotypic differences in eye colors, differential hair growth and coloring, "checkerboard" skin patterns, or missing or extraneous sexual organs. In 1998, at the University of Edinburgh, doctors examined a man who had complaints about an undescended left testicle. When they examined him, they were shocked to find an ovary and a fallopian tube in the male patient!

Most human chimeras, however, are not even aware of their conditions, because many of them appear completely normal. The most famous cases of chimerism to date are the linked cases of Lydia Fairchild (Year 2002) and Karen Keegan (Year 2000). Fairchild was pregnant with her third child, when she was separated from her partner, James Townsend. In order to obtain state welfare, she had to prove that she was the biological mother of her two born children. It was discovered, through DNA testing, that it was impossible that she was the biological mother of her two children because she bore no genetic similarity to them whatsoever. A case of welfare fraud ensued because the prosecutors believed the DNA results to be irrefutable. Even the testimony of Dr. Leonard Dreisbach, the obstetrician who had helped Fairchild give birth, did little to persuade the court in Fairchild's favor. The judge, perplexed by seemingly conflicting evidence, ordered that the third child, when born, to be tested as well. Surprisingly, the third child also showed no genetic similarities as well.

Fortunately for Fairchild, Karen Keegan also had similar experiences. Keegan needed a kidney transplant, and DNA testing for a compatible match with her two eldest sons showed that she had no genetic similarities to them at all. However, the doctors who worked with Keegan were familiar with the concept of chimerism and suggested that Keegan undergo further testing. Testing of her brothers and husband proved that her sons were related to them. Subsequent sampling of her skin and hair proved to be futile, but eventually matching DNA was found in her thyroid gland. It was the publication of this case, in the *New England Journal of Medicine*, which offered new insight on the case of Lydia Fairchild. Fairchild was found later on to be a chimera, with the second set of DNA found from her cervical smear. It was concluded that both Keegan and Fairchild were tetragametic.

Chapter Seven

2000 Year Old Analog Computer found

The Antikythera mechanism is an ancient mechanical computer designed to calculate astronomical positions. It was recovered in 1900–1901 from the Antikythera wreck. Its significance and complexity were not understood until decades later. Its time of construction is now estimated between 150 and 100 BC. The degree of mechanical sophistication is comparable to a 19th century Swiss clock. Technological artifacts of similar complexity and workmanship did not reappear until the 14th century, when mechanical astronomical clocks were built in Europe. Jacques-Yves Cousteau visited the wreck for the last time in

1978, but found no additional remains of the Antikythera mechanism. Professor Michael Edmunds of Cardiff University who led the most recent study of the mechanism said: "This device is just extraordinary, the only thing of its kind. The design is beautiful, the astronomy is exactly right. The way the mechanics are designed just makes your jaw drop. Whoever has done this has done it extremely carefully ... in terms of historic and scarcity value, I have to regard this mechanism as being more valuable than the Mona Lisa."

The device is displayed at the National Archaeological Museum of Athens, accompanied by a reconstruction made and donated to the museum by Derek de Solla Price. Other reconstructions are on display at the American Computer Museum in Bozeman, Montana, the Computer History Museum in Mountain View, California, the Children's Museum of Manhattan in New York, and in Kassel, Germany.

Function

The device is remarkable for the level of miniaturization and for the complexity of its parts, which is comparable to that of 19th-century clocks. It has more than 30 gears, although Michael Wright has suggested as many as 72 gears, with teeth formed through equilateral triangles. When a date was entered via a crank (now lost), the mechanism calculated the position of the Sun and Moon or other astronomical information such as the locations of planets. Since the purpose was to position astronomical bodies with respect to the celestial sphere, with reference to the observer's position on the surface of the Earth, the device was based on the geocentric model.

The mechanism has three main dials, one on the front, and two on the back. The front dial has two concentric scales. The outer ring is marked off with the days of the 365-day Egyptian calendar, or the Sothic year, based on the Sothic cycle. Inside this, there is a second dial marked with the Greek signs of the Zodiac and divided into degrees. The calendar dial can be moved to compensate for the effect of the extra quarter day in the solar year (there are 365.2422 days per year) by turning the scale backwards one day every four years. Note that the

Julian calendar, the first calendar of the region to contain leap years, was not introduced until about 46 BC, up to a century after the device was said to have been built.

The front dial probably carried at least three hands, one showing the date, and two others showing the positions of the Sun and the Moon. The Moon indicator is adjusted to show the first anomaly of the Moon's orbit. It is reasonable to suppose the Sun indicator had a similar adjustment, but any gearing for this mechanism (if it existed) has been lost. The front dial also includes a second mechanism with a spherical model of the Moon that displays the lunar phase.

There is reference in the inscriptions for the planets Mars and Venus, and it would have certainly been within the capabilities of the maker of this mechanism to include gearing to show their positions. There is some speculation that the mechanism may have had indicators for all the five planets known to the Greeks.

Source: http://en.wikipedia.org/wiki/Antikythera_mechanism

Chapter Eight

How did the Moon Form? Who Built the Moon?

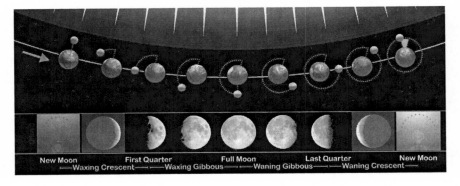

New Moon	First Quarter	Full Moon	Last Quarter	New Moon
——Waxing Crescent——	——Waxing Gibbous——	——Waning Gibbous——	——Waning Crescent——	

Not only is the Moon an apparently impossible object, we could not come to any other conclusion than the Moon is artificial. Is the Moon the creation of Intelligence?

The Moon is not only extremely odd in its construction, it also behaves in a way that is nothing less than miraculous. There are more than enough anomalies about the Moon to suggest it is not a naturally occurring body and was quite possibly engineered to sustain life on Earth.

The question of why the Moon had to be built is easy to answer. To produce life especially humans.

The Moon has some unique benefits for us humans. It has been nothing less than the incubator for life. If the Moon was not the exact mass, size and distance, that it has been at each stage of the Earth's evolution, there would be no intelligent life here. Scientists are agreed that we owe everything to the Moon. Without our Moon, the Earth would be as dead and solid as Venus,

In the book 'Who built the Moon" that brought to light some extraordinary facts about the Moon, the authors Christopher Knight and Alan Butler claimed that:

- It is exactly 400 times smaller than the Sun, but 400 times closer to the Earth, so that both the Moon and the Sun appear to be precisely the same size in the sky, which gives us the phenomenon we call a total eclipse. Whilst we take this for granted, it has been called the biggest coincidence in the universe.
- Something has to maintain the Moon with its precise altitude, course and speed for it to maintain orbit.
- The Moon sits very close to the Earth, yet it is widely regarded as the strangest object in the known universe.
- The Moon does not have a solid core like every other planetary object. It is either hollow or has a very low intensity interior. Bizarrely, its concentration of mass are located at a series of points just under its surface which caused havoc with early lunar spacecraft.
- It acts as a stabilizer that holds our planet at just the right angle to produce the seasons and keep water liquid across most of the planet.
- The moon is unlike any satellite in our entire universe. Because it is certain that it is 4.6 billion years old that raises some interesting points. The moon is older than the Earth by nearly 800,000 years according to scientific dating.
- Our Moon is the only moon in the solar system that has a stationary, near perfect circular orbit. It doesn't spin like a

natural celestial body. Our Moon shares no characteristics of any moon within our Solar System.

- From any point on the surface of Earth, only one side of the Moon is visible.

Dr. Sean C. Solomon of the Massachussetts Institute of Technology said the lunar orbiter experiments had vastly improved knowledge of the Moon's gravitational field, and indicated the "frightening possibility that the moon might be hollow".

Isaac Asimov, a Russian professor of biochemistry, stated "We cannot help but come to the conclusion that the moon by rights ought not be there. The fact that it is, is one of those strokes of luck almost too good to accept. Small planets such as Earth, with weak gravitational fields, might well lack satellites...In general, then, when a planet does have satellites, those satellites are much smaller than the planet itself. Therefore, even if the Earth has a satellite, there would be every reason to suspect, that at best, it would be a tiny world, perhaps, 30 miles in diameter, but that is not so. Earth not only has a satellite, but it is a giant satellite, 2160 miles in diameter."

Some lunar rocks have been found to contain 10 times more titanium than titanium rich rocks on Earth. Titanium is used in supersonic jets, deep diving submarines and spacecraft. Dr Harold Urey, a winner of the Nobel Prize for Chemistry, said he was "terribly puzzled by the rocks from the Moon, and in particular their Titanium content'. He said the samples were "mind-blowers", and that he could not account for the Titanium.

Irwin Shapiro from the Harvard-Smithsonian Center for astrophysics said, "The best explanation for the Moon is observational error---the Moon does not exist".

Robin Brett, a NASA scientist stated, "It seems easier to explain the non existence of the Moon than its existence."

BE SKEPTICAL EVEN OF ONE'S OWN SKEPTICISM

In November 1969, NASA intentionally crashed a lunar module causing an impact equivalent to one ton of TNT. The shockwaves built up and NASA scientists said the Moon "rang like a bell". The reverberation from it continued for 30 minutes.

Ken Johnson, a supervisor of the data and photo control department during the Apollo Missions said that the Moon not only rang like a bell, but the whole Moon "wobbled" in such a precise way, that it was "almost as though it had a gigantic hydraulic damper struts inside it.

In 1970, Mikhail Vasin and Alexander Shcerbakov from Soviet Academy of Sciences, produced an article for Sputnik magazine called "Is the Moon the creation of Alien Intelligence?" The article contained the following : The outer surface of the Moon is extremely hard and contains minerals like Titanium. Moon rocks have been found to contain processed metals, including Brass and Mica, and the elements Uranium 236 and Neptunium 237 that have never been found to occur naturally. Uranium 236 is a long lived radioactive nuclear waste and is found in spent nuclear and reprocessed Uranium. Neptunium 237 is a radioactive metallic element and a by product of nuclear reactors and the production of Plutonium. If a material had to be devised to protect a giant artificial satellite from the unfavorable effects of temperature, from cosmic radiation and meteorite bombardment, the experts would probably have hit upon these refractory metals. In that case, it is not clear why lunar rock is such an extraordinary poor heat conductor—a factor which so amazed astronauts ? Wasn't that what the designers of this super Sputnik of the Earth were after ? *From the engineer's point of view, this spaceship of ages long past which we call the Moon is superbly constructed"*.

Part 5

New Mentality Brings about New Discoveries

In Man's conquest of progress, his own prejudice and ignorance must be the first to surrender. From there, he will step his way across the heavens to the edge of infinity. No doubt he will stumble and fumble. Each step will be as uncertain as the last, yet each will bring him closer to ultimate truth. Throughout history, a handful of brave scientists and technicians pave the way to the future. Their mission: to collect information that will eventually enable man to step into the better world of tomorrow, to use their present knowhow as a springboard to the brave new world that they dare not imagine. They make their slow, uncharted way across the waters of mysticism, pseudoscience, and the paranormal of their times. The water's surface has depths and dangers that are, as yet, unprobed . . .

It is with these people that our new world has come to being, mapped, chartered, and indexed . . .

If we accept the tenuous fragility of modern progress through trial and error, misfortunes and sacrifice, perhaps we can begin the second dawning of humankind . . . today.

Chapter One

Monoatomic Gold as Superconductor

http://www.halexandria.org/dward479.htm

Dan Sewell Ward

A Monoatomic form of the element—in which each single atom is chemically inert and no longer possesses normal metallic characteristics; and in fact, may exhibit extraordinary properties. The atom's intrinsic temperature is now about one oK or close enough to absolute zero that superconductivity is a virtually automatic condition.

A case in point is gold. Normally a yellow metal with a precise electrical conductivity and other metallic characteristics, the metallic nature of gold begins to change as the individual gold atoms form chemical combinations of increasingly small numbers. At a microcluster stage, there might be thirteen atoms of gold in a single combination. Then, dramatically, at the monoatomic state, gold becomes a forest green color, with a distinctly different chemistry. Its electrical conductivity goes to zero even as its potential for superconductivity becomes maximized. Monoatomic gold can exhibit substantial variations in weight, as if it were no longer fully extant in space-time.

Other elements which have many of these same properties are the precious metals, which include ruthenium, rhodium, palladium, silver, osmium, iridium, platinum, and gold. All of these elements have, to a greater or lesser degree, the same progression as gold does in continuously reducing the number of atoms chemically connected. Many of these precious elements are found in the same ore deposits and in their monoatomic form are often referred to as the white powder of gold. Monoatomic elements apparently exist in nature in abundance. Precious metal ores are, however, not always assayed so as to identify them as such. Gold miners, for example, have found what they termed "ghost gold"—"stuff" that has the same chemistries as gold, but which were not yellow, did not exhibit normal electrical conductivity, and were not identifiable with ordinary emission spectroscopy. Thus they were more trouble than they were worth, and generally discounted.

The mining activity of what is considered the best deposit in the world for six of these elements (Pd, Pt, Os, Ru, Ir, and Rh) yields one third of one ounce of all these precious metals per ton of ore. But this is based on the standard spectroscopic analysis. The distinguishing characteristic between the first and second readings of the emission spectroscopy for the precious metals is that all of them come in two basic forms. The first is the traditional form of metals, yellow gold, for example. The second is the very nontraditional form of the metal, the monoatomic state. The chemistries and physics of these two different states of these metals are radically different. More importantly, when the atoms are in the monoatomic state, things really begin to get interesting!

A key to understanding monoatomic elements is to recognize that the monoatomic state results in a rearrangement of the electronic and nuclear orbits within the atom itself. This is the derivation of the term: Orbitally-Rearranged Monoatomic Element (ORME).

A monoatomic state implies a situation where an atom is "free from the influence of other atoms." Is this, perhaps, a violation of some very basic, absolutely fundamental law of the universe—which says that nothing is separate? If such a law constituted reality, then a necessary condition for monoatomic elements to even exist would require them to be superconductive, just in order to link them through all distance and time to other superconducting monoatomic elements. This would be necessary in order to prevent separation.

Chapter Two

Anticipation of Discovery in Astronomy Always Leads to New Discoveries

It was perturbations in Uranus's motion that led to the discovery of Neptune.

Astronomers had noticed increasing errors in their models for the orbit of Uranus and eventually ruled out many explanations besides an outer planet that could be perturbing its orbit.

In 1846, the planet Neptune was discovered after its existence was predicted.

This precise prediction of the new planet and its location was striking confirmation of the power of Newton's theory of gravitation.

Later, similar calculations on supposed perturbations of the orbits of Uranus and Neptune suggested the presence of yet another planet beyond the orbit of Neptune. Eventually, in 1930, a new planet, Pluto, was discovered.

We have even more amazing discoveries recently:

A trans-Neptunian object is any object in the solar system that orbits the sun at a greater average distance than Neptune.

Sedna

90377 Sedna is a large trans-Neptunian object, which, as of 2012, was about three times as far from the sun as Neptune. For most of its orbit, it is even farther from the sun than at present, with its aphelion estimated at 937 astronomical units, making it one of the most distant known objects in the solar system other than long period comets. Sedna's exceptionally long and elongated orbit, taking approximately 11,400 years to complete, and distant point of closest approach to the sun, at 76 AU, have led to much speculation as to its origin.

Eris, Pluto, Makemake, and Haumea

As of November 2009, two hundred trans-Neptunian objects have their orbits well enough determined that they have been given a permanent minor planet designation.

The largest known trans-Neptunian objects are Eris and Pluto, followed by Makemake and Haumea. The Kuiper belt, scattered disk, and Oort cloud are three conventional divisions of this volume of space.

Comet Lovejoy

Comet Lovejoy is a long period comet and Kreutz Sungrazer. It was discovered in November 2011 by Australian amateur astronomer Terry Lovejoy. The comet's perihelion took it through the sun's corona on 16 December 2011, after which it emerged intact and continued on its orbit to the outer solar system. In the process, it astounded all scientists.

Comet 96P/Machholz 1

Comet 96P/Machholz last came to perihelion on July 14, 2012 and will next come to perihelion on October 27, 2017. 96P/Machholz has an estimated radius of around 3.2 km.

Machholz 1 is unusual among comets in several respects. Its highly eccentric 5.2-year orbit has the smallest perihelion distance known among numbered/regular short period comets, bringing it considerably closer to the sun than the orbit of Mercury.

Chapter Three

God Particle Having Been in Existence for Millenia, Found in July 2012

http://www.technewsworld.com/story/77540.html

By Richard Adhikari TechNewsWorld 03/15/13

"There is no indication that this is not a Standard Model Higgs at present," Dan Green, a senior scientist emeritus at Fermilab's CMS Center, told TechNewsWorld. "More data has shrunk the possibilities of a non-Standard Model Higgs."

A large decay rate into two photons was seen earlier, but that had large statistical errors, and the data "has now settled to a value closer to the Standard Model," he said.

Next Steps

The issue scientists are facing now is finding out what protects the Higgs boson's low mass from radiative corrections. "Something limits the mass to about 125 GeV and we don't know what that is—yet," Green said.

There's a level of certainty that the particle discovered is indeed the Higgs boson, but "in science, you cannot say with certainty what you have seen," he said. The scientists want what they refer to as "5-sigma" confidence "for all the possible decay modes of the Higgs, and that will take several years of the LHC running at increased luminosity and higher energy."

The scientists are "being super-cautious," William Newman, professor of Earth and space sciences at the University of California at Los Angeles, told TechNewsWorld. "The original discovery was a 5-sigma discovery and that's a super-gold standard; it's almost the holy grail."

Validation of the standard model at this level "excludes a large variety of competing models that people have been putting forth, and that dramatically constrains how the universe can behave," he said.

Why Search for Decay?

The Higgs boson has no spin, no electric charge and no color charge, and is its own antiparticle. It doesn't exist except as a theoretical construct, and here's why: It accompanies the Higgs Field, a possible invisible field of energy that exists throughout the universe.

In the standard model, the Higgs Field consists of four components: two charged component fields and two neutral ones. Both the charged components and one of the neutral fields are Goldstone bosons. The quantum of the last, neutral component is theoretically realized as the Higgs boson.

Quantum mechanics theory holds that particles will decay into a set of lighter particles if it's possible for them to do so. The likelihood of this occurring depends on various factors, including the difference in mass between the original particle, and the ones it decays into, and the strength of the interactions. The standard model fixes most of these factors except for the mass of the Higgs boson itself.

Chapter Four

Galileo, Your Best Bet Is to Be Born after Newton

Sir Isaac Newton, born on December 25, 1642, was an English physicist, mathematician, astronomer, natural philosopher, alchemist, and theologian who has been considered by many to be the greatest and most influential scientist who ever lived.

Galileo Galilei, born on February 15, 1564, was an Italian scientist who supported Copernicanism, the idea that Earth orbits the sun. Galileo defended his views in *Dialogue Concerning the Two Chief World Systems*. For doing so, he was tried by the Roman Inquisition, was found "suspect of heresy," and spent the rest of his life under house arrest. His findings changed our world view for all time. Galileo died a natural death in the year 1642. He was seventy-seven years old and was serving his life imprisonment sentence in Arcetri when he died.

In fact, Newton did not discover gravity. It was actually Galileo. Galileo dropped different weight objects from the Tower of Pisa and noted they all fell at the same rate, and he did many other experiments along these lines.

In fact, Galileo died a year before Newton was born.

What if Galileo Galilei were born after Sir Isaac Newton?

Galileo Galilei would have had an easier life if he were born after Sir Isaac Newton. With the discovery of gravitational pull, it could be demonstrated easily that the Earth rotated around the Sun. And he wouldn't have spent the rest of his life under house arrest.

Sometimes, some truths or discoveries come too early, especially for people who are indifferent, arrogant, and ignorant. Politics, greed, and an irrational resistance to change have prompted people to denounce the latest discovery or technical innovation as rubbish, often as a knee-jerk reaction. Egos, vested interests, politics, moral and religious objections, and evasion of responsibility are all underlying factors to the fact that many beneficial discoveries have been swept under the carpet.

Many, many lives that could be saved with the latest discoveries and inventions have been lost with the latest findings suppressed and the inventor dying unheralded. Inventors often need a great deal of tenacity, as well as vision, to overcome the skepticism and ridicule that come as an avalanche. Sometimes discoverers and inventors are persecuted. A breakthrough would have far-reaching consequences for the entire religious sector based on Galileo Galilei's work—which, in itself, may be the problem. Until a very long time after his death, his books were banned and his writings and findings were considered deviant, false and of the devil.

Now who is to say what discoveries should come earlier than others?

Conclusion

Sitchin may be a hoax in some people's eyes. We might not value what he has to offer us now. In time, when other scientific discoveries are made, they may exonerate Sitchin. To Sitchin's critics, who would like to see his books banned and his writings and findings considered deviant and false, they really need to appreciate how Sitchin's findings have changed our world view for all time.

Chapter Five

The Greeks Knew the Earth Was Round and Measured Its Circumference Before Copernicus and Galileo

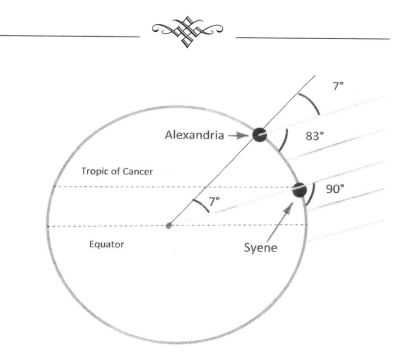

The ancient Greeks knew a thing or two. Nowadays we say Copernicus discovered the Earth was round, for example, but the Greeks had that figured out almost two thousand years earlier!

How Did They Know the Earth Was Round?

The common Greek knew the Earth was round back then because Greeks where sailors. As a ship would go away from a port they noticed that the sails of the ship would disappear last. Thus they knew that the Earth has a curvature and was not flat!

Now remember that all the stars seem to move in a big circle over our heads because the Earth itself is spinning. The North Star is the center of the circle, and stays put.

However, Aristotle noticed that when you are down south in Egypt, the North Star is close to the horizon. When you are up north in Greece again, it's high in the sky.

How is that possible, since the North Star is supposed to stay put? It's possible, Aristotle concluded, only if the Earth is round.

Think of it this way . . . When you are standing at the north pole, where is the North Star? Directly overhead. Now place yourself at the equator. Where is the North Star? Yep—it's down on the horizon, because you have traveled along a curve. Go even further south and you won't be able to see it at all; the Earth will hide it. Go further north and it seems to rise higher and higher.

What about Measuring the Earth's Circumference?

Contrary to popular belief, Christopher Columbus did not discover that the Earth is round. Eratosthenes (276–194 B.C.) made that discovery about seventeen hundred years before Columbus!

He calculated the circumference of the Earth without leaving Egypt.

Eratosthenes was a Greek mathematician, astronomer, and the father of geography with the invention of a system of latitude and longitude.

How did he do it?

1. Eratosthenes first measured the distance between Alexandria and Syene to be five hundred miles. This is based on the estimated average speed of a caravan of camels that traveled this distance. Camels traveled the distance many times to obtain an average estimate of the distance.

2. Eratosthenes observed that at noon on the summer solstice, the longest day of the year, no shadow was cast in a well in the city of Syene in Egypt.

3. At that time the sun was directly overhead. He then assumed that the sun was so far off that its rays hit the Earth in parallel.

4. He further assumed that the Earth was shaped like a ball from the stories of Greek sailors.

5. He surveyed the city of Syene to be on the same meridian as the city of Alexandria. because the sun set at the same time when it was directly between the two cities.

6. So, on the summer solstice, at noon, in Alexandria, Eratosthenes measured the angle of the sun's rays to be 7.2 degrees angle with a tall rod.

7. He also knew, from measurement, that in his hometown of Alexandria the angle of elevation of the sun was 1/50th of a circle (7°12') south of the zenith on the solstice noon.

8. Assuming that the Earth was spherical (360°), and that Alexandria was due north of Syene, he concluded that the meridian arc distance from Alexandria to Syene must therefore be $1/50 = 7°12'/360°$, and was therefore 1/50 of the total circumference of the Earth.

9. Since he knew that five hundred miles was 1/50 of the circumference of the Earth, he needed to use fifty of these lengths to surround the Earth.

10. So he multiplied five hundred miles by fifty to get twenty-five thousand miles.

11. Then he added two hundred miles more to make up for what he thought were bad measurements.

So he calculated the circumference of the Earth to be 25, 200 miles.

That measurement is close to the actual 24,901 miles. In those days they couldn't easily make very accurate measurements of the distance between two places, so this could cause a lot of error.

Chapter Six

"Hi, Roman Citizens . . . Rome Is Not the Only Mighty Empire in the World!"

"You called yourself the greatest empire in existence. Do you believe there exists an empire called the Chinese Empire, which is at least as large if not bigger than the Roman Empire?"

"You are most proud of your architectural achievements. Do you know there are pyramids that are very much harder to construct?"

"Other than the Romans and barbarians, do you know of the existence of the Eskimos, American Indians, Australian Aborigines . . . ?"

"Do you believe your Roman scholars and authorities know most of everything? Do you have many financial crises? Social Inequalities?"

"Would you believe in some strange lands, strange people, strange ideas, and strange discoveries if your Roman scholars and authorities told you not to?"

"Do you know Korea? The country has been in existence for thousands of years!"

Well, the last question is too hard. Many Americans did not even know Korea existed until the American government decided to send troops to Korea in the Korean War, despite the fact that the country has been in existence for thousands of years!

With these facts in mind, how are we even sure on what we know, or for that matter, what we don't know?

Conclusion

Sitchin was trying hard to tell us something we don't know.

Chapter Seven

Conversation with the Wright Brothers in 1902, Just before Their Important Breakthrough

If we go back to 1902, with all the knowledge we have about flying and aerodynamics, will we find any person who can be convinced that flying is possible, let alone who can be convinced that man would be on the brink of flying through the air at the speed of sound? Can we even convince the Wright Brothers? Maybe not.

Wright Brothers: Well, Mr. Flying, if your motorized craft can go faster than the speed of sound, why isn't your flimsy craft torn to pieces by the wind? Besides, the human body cannot take strong acceleration, according to the laws of nature.

Mr. Flying: It is possible to pass the sound barrier by designing the wings and body to move the shock wave down the plane as you surpass the speed of sound.

Wright Brothers: How do you make wings with enough lifting power? How do you avoid the airplane spinning out of control by the wing warping mechanism?

Mr. Flying: Combinations of aluminum, magnesium, and small amounts of copper and manganese make a light but strong alloy called duralumin, which is suitable for airplane parts.

Wright Brothers: Really? Planes made of alloy? Well, if your plane can fly faster than sound, then why don't you just fly to the moon?

Mr. Flying: Trained astronauts in a space shuttle can do that, not an airplane.

Wright Brothers: I see. Fly to the moon in a space shuttle? Man will probably not fly in their lifetime. Not in an airplane or space shuttle. Is that a fantasy?

Mr. Flying: Perhaps, perhaps not.

Chapter Eight

Suppressed Discoveries Become Forbidden and Then Forgotten

Author Jonathan Eisen has collected over forty intriguing stories of scientific cover-ups and programs of misinformation concocted to conceal some of the most phenomenal innovations in mankind's history. In his book, *Suppressed Inventions and Other Discoveries*, he presents documented evidence that corporate self-interest, scientific arrogance, and political savvy have contrived to keep us in the dark about technological breakthroughs or interplanetary contact that may shift the current balance of power.

Area I

- Alternative Medical Therapies
- The Great Fluoridation Hoax
- Deadly Mercury—How It Became Your Dentist's Darling
- The Alzheimer's Cover-Up
- Vaccinations—Adverse Reactions Cover-Up
- AIDS and Ebola—Where Did They Really Come From?
- Polio Vaccines and the Origin of AIDS
- Oxygen Therapies, the Virus Destroyers
- Oxygen Therapy
- The FDA
- Harry Hoxsey—The AMA's Successful Attempt to Suppress My Cure for Cancer
- Royal Raymond Rife and the Cancer Cure That Worked!
- The Persecution and Trial of Gaston Naessens

- Dr. Max Gerson's Nutritional Therapy for Cancer and Other Diseases

Area II

- The Suppression of Unorthodox Science
- Science as Credo
- Sigmund Freud and the Cover-Up of "The Aetiology of Hysteria"
- The Burial of Living Technology
- Egyptian History and Cosmic Catastrophe
- Archaeological Cover-Ups?
- Introduction to Bread From Stones
- Scientist With an Attitude—Wilhelm Reich
- The AMA's Charge on the Light Brigade
- The Neurophone

Area III

- The Suppression of UFO Technologies and Extraterrestrial Contact
- Breakthrough as Boffins Beat Gravity
- Antigravity on the Rocks—The T. T. Brown Story
- Did NASA Sabotage Its Own Space Capsule?
- Extraterrestrial Exposure Law Already Passed by Congress
- The Stonewalling of High Strangeness
- UFOs and the U.S. Air Force
- UFOs and the CIA—Anatomy of a Cover-up
- NASA
- UFO Phenomena and the Self-Censorship of Science
- Mars—The Telescopic Evidence
- Never a Straight Answer—A Book Review of NASA Mooned America

Area IV

- The Suppression of Fuel Savers and Alternate Energy Resources
- Nikola Tesla—A Brief Introduction
- Tesla's Controversial Life and Death
- Transmission of Electrical Energy without Wires
- From the Archives of Lester J. Hendershot
- Gunfire in the Laboratory—T. Henry Moray and the Free Energy Machine
- Sunbeams from Cucumbers
- Archie Blue
- The Story of Francisco Pacheco
- Amazing Locomotion and Energy Super Technology and Carburetors
- The Charles Pogue Story
- News Clips on Suppressed Fuel Savers

Chapter Nine

The Elephant of Reality: God, Evolution, and Intelligent Design

Chronology of Human Events and the Rise of Different Theories of Human Origins

4.5 Billion Years Ago

Our sun created the planet Tiamet, the proto-Earth. Tiamat orbited the Sun counterclockwise.

Two Hundred Million Years Ago

Panspermia started two hundred million years ago when Nibiru's moon and Nibiru hit Tiamat and left no crust at all in the Pacific Gap, only a gaping hole. Initial life-forms were transferred to New Earth.

Evolution slowly picked up pace as conditions changed for the better to favor the emergence of life-forms.

Four Hundred and Fifty Thousand Years Ago

Annunaki humanoids arrived on planet Earth, setting up Nibiru Colony

Four Hundred Thousand Years Ago

The Annunaki started **non-intelligent design.** They created many hybrid species and chimeras out of genetic errors. They also created many abhorrent types of experimental humanoids and animals.

Three Hundred Thousand Years Ago

The Annunaki started **intelligent design.** They created the perfected specimens of Adam and Eve as human progenitors to be used as mining workers.

Two Hundred and Fifty Thousand Years Ago

Humans multiplied in great numbers. For effective rule of the human population, the Annunaki humanoids taught the humans to revere them as **gods.**

12,000 B.C.

The Annunaki strengthened their lordship over the humans. Various structures for worshipping them as **gods** were erected. Treating humans as domesticated animals, they brought new species of plants and animals from their home planet Nibiru as rewards for humans who labored in the mines. They also domesticated wild plants and animals native to Earth for humans as food and other consumption.

9,000 B.C

Intermarrying earthlings, especially females, further hybridized the human genome. Men of renown known as heros or nephilims further sped up **human evolution.** Human beings now seemed to be perfectly designed with a purpose. This prompted posterity to term their birth through "**intelligent design.**"

6,000 B.C.

The Anunnaki created bloodlines to rule humanity on their behalf, and these are the families still in control of the world to this day. Kingship was granted to humanity by the Anunnaki, and it was originally known as Anuship after An or Anu, the ruler of the 'gods.' Eventually, kingships became prevalent all over the globe. The royal families and aristocracy of Europe, Asia, and the Middle East are obvious examples of this. Most of the world's organized religions were also established at around this time.

5,000 B.C.

The constant power struggle and the quest for domination by the Annunaki led a minority of humans to escape the control of the Annunaki by migrating to other areas. Some of these became nonbelievers of the Annunaki. Their teachings were passed down to posterity. Some of their descendants became **agnostics and atheists.**

1,831 A.D.

"The Voyage of the Beagle" refers to the second survey expedition of the ship HMS Beagle, which set sail from Plymouth Sound on 27 December 1831 under the command of Captain Robert FitzRoy, R.N. Darwin's notes made during the voyage included comments illustrating his changing views at a time when he was developing his theory of Evolution by Natural Selection and included some suggestions of his ideas, particularly in the second edition of 1845.

1855 A.D.

In February 1855, while working in Sarawak on the island of Borneo, Alfred Russel Wallace wrote "On the Law which has Regulated the Introduction of New Species", a paper which was published in the Annals and Magazine of Natural History in September 1855. By February 1858, Wallace had been convinced by his biogeographical research in the Malay Archipelago of the reality of evolution.

1859 A.D.

On November 1859, Charles Darwin published "On the Origin of Species", a work of scientific literature by which is considered to be the foundation of evolutionary biology.

2013 A.D.

History replays itself over and over again in man's history of conflicts, conquests, and conjectures. Man has been warring one another as individuals and as nations. All theories of human origins seek to dominate to the exclusion of other theories.

Part 6

Pursue Truth Aggressively—Your Truth, Especially

We have begun the space age. We are all trying to maneuver through a brand-new, unfamiliar landscape using an old map drawn up by religion and Darwin. We need a new guiding principle.

Unaware of the new realities, we are like the last remaining jungle fighters of Japan still hiding in the swamps blindly unaware that the war was over decades ago.

"The past history of humanity shows us that each stage of its development necessitates an uprooting and renewing of fundamental beliefs in our scientific, social, philosophic, and religious conceptions."—M. Plank. Our history is also full of silly mistakes that must make us smile today. Just like new things that happen everyday, another new theory will soon appear, and the new theory that just appeared will be replaced by a newer one, which itself will be replaced by a better one, and so on ad infinitum . . . Our history is also full of icons who also made silly mistakes.

Ernest Rutherford implied that energy produced by an atom was insignificant and anyone who expected to find a new source of energy in that transformation was dreaming. Even Albert Einstein indicated that there was no indication that nuclear energy would be accessible one day, and yet Hiroshima happened twelve years later.

Author Arthur C. Clarke believed that when even eminent and distinguished scientists say something is possible, he could be right, but if he says it's impossible, then there's a good chance he's wrong.

Sitchin's work will make us dream impossible dreams that we have not dreamed before. With that, he will bring us to new progress that we cannot even dream of, if we understand our past and then foresee our future. It is a new reality too incredible to accept, too awesome to face. The establishment of a new paradigm need not cause the fall of the old paradigm. The time has come to shift into a higher gear and reveal what the future holds for us by examining our very own past.

God particles are around for billions of years but we only know they exist in 2012!

Over the years, startling evidence has been uncovered, challenging established notions of the origins of life on Earth—evidence that suggests the existence of an advanced group of extraterrestrials who once inhabited our world.

Many researchers have since uncovered incredible findings having depth, complexity, and far-reaching effects in support of Sitchin. This book is an attempt to bring about the general awareness in the public of a topic so important and so controversial. The reader is encouraged to dig deeper into all the material, the artifacts, the "what if," and reach his or her own conclusions.

We need to be aware of our own tunnel vision imposed by the current scientific and religious orthodoxy. "Your time is limited, so don't waste it living someone else's life. Don't be trapped by dogma—which is living with the results of other people's thinking. Don't let the noise of others' opinions drown out your own inner voice. And most important, have the courage to follow your heart and intuition. They somehow already know what you truly want to become. Everything else is secondary" (Stanford commencement speech, June 2005, Steve Jobs).

Wake up! This is not science fiction. Truths that were yet unknown to this day have been unveiled.

The highest form of ignorance is when you reject something you don't know anything about.

Wayne Dyer

"What we call imagination is actually the universal library of what's real. You couldn't imagine it if it weren't real somewhere, sometime.

~ Terence Mckenna ~

Review Requested:

If you loved this book, would you please provide
a review at Amazon.com?
You can also reach the author at info@elvisnewman.com

CPSIA information can be obtained at www.ICGtesting.com
Printed in the USA
LVOW07s0746280216

476958LV00001BA/179/P